POLITICS AND ECONOMICS OF THE MIDDLE EAST

IRAN'S BALLISTIC MISSILE AND SPACE LAUNCH PROGRAMS

ELEMENTS AND CONSIDERATIONS

POLITICS AND ECONOMICS OF THE MIDDLE EAST

Additional books in this series can be found on Nova's website under the Series tab.

Additional E-books in this series can be found on Nova's website under the E-book tab.

GLOBAL POLITICAL STUDIES

Additional books in this series can be found on Nova's website under the Series tab.

Additional E-books in this series can be found on Nova's website under the E-book tab.

POLITICS AND ECONOMICS OF THE MIDDLE EAST

IRAN'S BALLISTIC MISSILE AND SPACE LAUNCH PROGRAMS

ELEMENTS AND CONSIDERATIONS

STEVEN WIETMAN
EDITOR

nova publishers
New York

For permission to use material from this book please contact us:
Telephone 631-231-7269; Fax 631-231-8175
Web Site: http://www.novapublishers.com

Library of Congress Cataloging-in-Publication Data

ISBN: 978-1-62417-988-4

Published by Nova Science Publishers, Inc. † New York

CONTENTS

Preface		**vii**
Chapter 1	Iran's Ballistic Missile and Space Launch Programs *Steven A. Hildreth*	**1**
Chapter 2	Iran: U.S. Concerns and Policy Responses *Kenneth Katzman*	**63**
Index		**145**

PREFACE

Why does Iranian acquisition of ballistic missiles or Iran's pursuit of a space launch capability matter, especially to the Congress? For decades, most in Congress have viewed Iran with concern because of its nuclear program and its support of anti-U.S. and anti-Israel militant movements in the region. Although it is not certain that Iran has made the decision to develop a nuclear weapon, it is taking steps to drastically reduce the time needed to obtain nuclear weapons should a decision be made to do so. It is the prospect of an Iranian nuclear weapon mated to an effective missile delivery capability that is especially worrisome to most. This book examines the current and prospective Iranian capabilities that challenge U.S. national security interests, with a focus on its ballistic missile and space launch programs.

Chapter 1 – Iran has long been a source of concern for the United States and other countries because its goals are at odds with core U.S. objectives in the Middle East. Although it is not certain that Iran has made the decision to develop a nuclear weapon, it is taking steps to drastically reduce the time needed to obtain nuclear weapons should a decision be made to do so. It is the prospect of an Iranian nuclear weapon mated to an effective missile delivery capability that is especially worrisome to most. Congress has long been interested in these matters. Congress has held numerous hearings on Iran, passed various resolutions regarding Iran and approved a range of sanctions against Iran over the past several decades. According to the U.S. government, Iran has the largest number of ballistic missiles in the Middle East; it is developing missiles and space launch vehicles for multiple purposes. Iran is pursuing its missile and space programs with development and testing facilities that are scattered throughout the country. Assessing Iran's ballistic missile programs is challenging for many reasons, including the lack of specificity in official public sources, the secretive nature of Iran's regime and the regime's frequent exaggerations of its ballistic missile capabilities, and the overwhelming amount of and often conflicting information found in non-official sources. The vast majority of Iran's heavy artillery rockets and ballistic missiles are short-range of less than about 500 kilometers. Most of Iran's ballistic missiles in fact are Scud-B and Scud-Cs, with a majority likely being Scud Cs, which are 500 km range capable. Iran views its short-range ballistic missiles (SRBM) capability as necessary for battlefield and tactical military purposes. These missiles could not strike U.S. or allied bases in the region unless they were moved far from their operating base and launched from vulnerable positions along Iran's Persian Gulf coastline. This is not likely because of logistical and operational security reasons. Although these SRBMs are not very accurate, they could be fired against economic or civilian targets. Also, any such missile attacks against U.S. bases, while not militarily decisive, could disrupt or complicate (but not halt) base operations. Iran has grown increasingly self-sufficient in the production of SRBMs, but it still probably relies on others for some key components. Gaining access to these kinds of critical components and materials has grown increasingly difficult for Iran. Stricter international enforcement of export controls and broadening sanctions have reportedly slowed down Iran's efforts and forced Iran to find less reliable alternative sources of rocket and missile technology. Iran is developing and producing medium-range ballistic missile (MRBM) capabilities with ranges estimated up to about 2,000 kilometers (with some non-U.S. government sources citing slightly higher ranges), sufficient to strike targets throughout the Middle East. U.S. intelligence assessments state such missiles are inherently capable of carrying a nuclear warhead. Although the number of Iran's MRBMs is thought to be relatively small by official U.S. estimates, it is expected to continue to build more capable MRBMs. Iran views these missiles as an important deterrent and retaliatory force against U.S. and other forces in the region in the event of war. Iran has also constructed an underground network of bunkers and underground silo-like missile launch facilities, and is seeking improved air defenses presumably to enhance the survivability of their MRBMs against preemptive attack. Currently Iran must rely on others for certain key missile components and materials in its MRBM program. Export controls and sanctions have made it increasingly difficult, but certainly not impossible, for Iran to acquire the best of such items. On the other hand, these export control measures and sanctions have forced Iran to try to exploit weaknesses in existing export and nonproliferation regimes, including by trying to find foreign sellers willing to circumvent those laws. Iran also has a genuine and ambitious space launch program, which seeks to enhance Iran's national pride, and perhaps more importantly, its international reputation as a growing advanced industrial power. Iran also sees itself as a potential leader in the Middle East offering space launch and satellite services. Iran has stated it plans to use future launchers for placing intelligence gathering satellites into orbit, although such a capability is a decade or so in the future. Many believe Iran's space launch program could mask the development of an intercontinental ballistic missile (ICBM) – with ranges in excess of 5,500 km that could threaten targets throughout Europe, and even the United States if Iran achieved an ICBM capability of at least 10,000 km. ICBMs share many similar technologies and processes inherent in a space launch program, but it seems clear that Iran has a dedicated space launch effort and it is not simply a cover for ICBM development. Since 1999, the U.S. Intelligence Community (IC) has assessed that Iran could test an ICBM by 2015 with sufficient foreign assistance, especially from a country such as China or Russia (whose support has reportedly diminished over the past decade). It is increasingly uncertain whether Iran will be able to achieve an ICBM capability by 2015 for several reasons: Iran does not appear to be receiving the degree of foreign support many believe would be necessary, Iran has found it increasingly difficult to acquire certain critical components and materials because of sanctions, and Iran has not demonstrated the kind of flight test program many view as necessary to produce an ICBM.

Chapter 2 – A priority of Obama Administration policy has been to address the perceived threat posed to a broad range of U.S. interests by Iran, in particular by its advancing nuclear enrichment program. These advances have caused the government of Israel to assert that it might take unilateral military action against Iran's nuclear facilities unless the United States provides assurances that it will act, militarily if necessary, to prevent Iran from taking the final steps toward developing a nuclear weapon. Aside from the nuclear issue, the United States has long seen Iran's support for militant groups in the Middle East and in Iraq and

Afghanistan as efforts to undermine U.S. interests and allies. U.S. officials also accuse Iran of helping Syria's leadership try to defeat a growing popular opposition movement and of taking advantage of Shiite majority unrest against the Sunni-led, pro-U.S. government of Bahrain. The Obama Administration has orchestrated broad international pressure on Iran through economic sanctions, while also offering Iran sustained engagement if it verifiably demonstrates to the international community that its nuclear program is peaceful. Iran's leaders returned to nuclear talks with six powers in April 2012 after a one year hiatus. Three rounds of talks held in April, in May, and in June yielded no breakthroughs, but did explore a potential compromise under which Iran might end uranium enrichment to 20% purity (a level not technically far from weapons grade) in exchange for some sanctions relief. Subsequent technical talks produced no immediate progress, but might have set the stage for more high level talks in late 2012 or early 2013. The Administration asserts that there is time for accumulated international sanctions to compel Iran's leaders to compromise before U.S. military action be considered. Since the beginning of 2012, as significant multilateral sanctions have been added on Iran's oil exports— including an oil purchase embargo by the European Union that went into full effect on July 1, 2012—the regime has begun to acknowledge significant economic pressure. Those pressure led to a virtual collapse in the market value of Iran's currency, the rial, in early October 2012. Since late 2011, both Britain and Canada, have closed their embassies in Iran. Some reports suggest that the Administration might consider bilateral talks with Iran if doing so could head off confrontation. The Administration and many outside experts also perceive that the legitimacy and popularity of Iran's regime is in decline, although not to the point where the regime's grip on power is threatened. There are few signs that the opposition in Iran or in exile have gained traction against the regime, even though international sanctions are causing clear public frustration. The reformist boycott of the March 2, 2012, parliamentary elections rendered the election a contest between factions supporting either President Mahmoud Ahmadinejad or Supreme Leader Ali Khamene'i. Khamene'i supporters were elected overwhelmingly. It is likely that only hardliners will be significant candidates in the next presidential election to be held on June 14, 2013. The 112th Congress has supported additional economic sanctions against Iran, most recently with enactment of the Iran Threat Reduction and Syria Human Rights Act of 2012 (H.R. 1905, P.L. 112-158), which expands sanctions against companies that conduct energy and financial transactions with Iran. Additional sanctions were imposed by the European Union in mid-October and some Members of Congress reportedly are considering legislation that would compel a near- total global trade embargo on Iran.

In: Iran's Ballistic Missile and Space Launch Programs
Editor: Steven Wietman
ISBN: 978-1-62417-988-4

Chapter 1

IRAN'S BALLISTIC MISSILE AND SPACE LAUNCH PROGRAMS*

Steven A. Hildreth

SUMMARY

Iran has long been a source of concern for the United States and other countries because its goals are at odds with core U.S. objectives in the Middle East. Although it is not certain that Iran has made the decision to develop a nuclear weapon, it is taking steps to drastically reduce the time needed to obtain nuclear weapons should a decision be made to do so. It is the prospect of an Iranian nuclear weapon mated to an effective missile delivery capability that is especially worrisome to most.

Congress has long been interested in these matters. Congress has held numerous hearings on Iran, passed various resolutions regarding Iran and approved a range of sanctions against Iran over the past several decades.

According to the U.S. government, Iran has the largest number of ballistic missiles in the Middle East; it is developing missiles and space launch vehicles for multiple purposes. Iran is pursuing its missile and space programs with development and testing facilities that are scattered throughout the country. Assessing Iran's ballistic missile programs is challenging for many reasons, including the lack of specificity in official public sources, the secretive nature of Iran's regime and the regime's frequent exaggerations of its ballistic missile capabilities, and the overwhelming amount of and often conflicting information found in non-official sources.

The vast majority of Iran's heavy artillery rockets and ballistic missiles are short-range of less than about 500 kilometers. Most of Iran's ballistic missiles in fact are Scud-B and Scud-Cs, with a majority likely being Scud Cs, which are 500 km range capable. Iran views its short-range ballistic missiles (SRBM) capability as necessary for battlefield and tactical military purposes. These missiles could not strike U.S. or allied bases in the region unless they were moved far from their operating base and launched from vulnerable positions along Iran's Persian Gulf coastline. This is not likely because of logistical and operational security reasons. Although these SRBMs are not very accurate, they could be fired against economic or civilian targets. Also, any such missile attacks

* This is an edited, reformatted and augmented version of Congressional Research Service, Publication No. R42849, dated December 6, 2012.

against U.S. bases, while not militarily decisive, could disrupt or complicate (but not halt) base operations.

Iran has grown increasingly self-sufficient in the production of SRBMs, but it still probably relies on others for some key components. Gaining access to these kinds of critical components and materials has grown increasingly difficult for Iran. Stricter international enforcement of export controls and broadening sanctions have reportedly slowed down Iran's efforts and forced Iran to find less reliable alternative sources of rocket and missile technology.

Iran is developing and producing medium-range ballistic missile (MRBM) capabilities with ranges estimated up to about 2,000 kilometers (with some non-U.S. government sources citing slightly higher ranges), sufficient to strike targets throughout the Middle East. U.S. intelligence assessments state such missiles are inherently capable of carrying a nuclear warhead. Although the number of Iran's MRBMs is thought to be relatively small by official U.S. estimates, it is expected to continue to build more capable MRBMs. Iran views these missiles as an important deterrent and retaliatory force against U.S. and other forces in the region in the event of war. Iran has also constructed an underground network of bunkers and underground silo-like missile launch facilities, and is seeking improved air defenses presumably to enhance the survivability of their MRBMs against preemptive attack. Currently Iran must rely on others for certain key missile components and materials in its MRBM program. Export controls and sanctions have made it increasingly difficult, but certainly not impossible, for Iran to acquire the best of such items. On the other hand, these export control measures and sanctions have forced Iran to try to exploit weaknesses in existing export and nonproliferation regimes, including by trying to find foreign sellers willing to circumvent those laws.

Iran also has a genuine and ambitious space launch program, which seeks to enhance Iran's national pride, and perhaps more importantly, its international reputation as a growing advanced industrial power. Iran also sees itself as a potential leader in the Middle East offering space launch and satellite services. Iran has stated it plans to use future launchers for placing intelligence gathering satellites into orbit, although such a capability is a decade or so in the future. Many believe Iran's space launch program could mask the development of an intercontinental ballistic missile (ICBM) – with ranges in excess of 5,500 km that could threaten targets throughout Europe, and even the United States if Iran achieved an ICBM capability of at least 10,000 km. ICBMs share many similar technologies and processes inherent in a space launch program, but it seems clear that Iran has a dedicated space launch effort and it is not simply a cover for ICBM development. Since 1999, the U.S. Intelligence Community (IC) has assessed that Iran could test an ICBM by 2015 with sufficient foreign assistance, especially from a country such as China or Russia (whose support has reportedly diminished over the past decade). It is increasingly uncertain whether Iran will be able to achieve an ICBM capability by 2015 for several reasons: Iran does not appear to be receiving the degree of foreign support many believe would be necessary, Iran has found it increasingly difficult to acquire certain critical components and materials because of sanctions, and Iran has not demonstrated the kind of flight test program many view as necessary to produce an ICBM.

CONGRESSIONAL INTEREST

Why does Iranian acquisition of ballistic missiles[1] or Iran's pursuit of a space launch capability matter, especially to the Congress? For decades, most in Congress have viewed Iran with concern because of its nuclear program and its support of anti-U.S. and anti-Israel militant movements in the region. Congress has held numerous hearings on Iran, passed various resolutions regarding Iran and approved a range of sanctions against Iran over the past several decades. Congress has established a broad number of reporting requirements of the Executive Branch that encompass the range of its concerns regarding Iranian WMD and ballistic missile programs. These reporting requirements are detailed in Appendix B, "Reporting Requirements Relevant to Iranian Ballistic Missile Programs." Congress has also supported funding BMD programs aimed at countering ballistic missiles especially those from Iran,[2] including cooperative programs with friendly and allied nations such as Israel. Iran's current and prospective capabilities challenge U.S. national security interests in many ways and are raised briefly below.

In response to the Iranian ballistic missile and space program, Congress is considering the deployment of an ballistic missile defense site on the East coast to defend against possible Iranian ICBMs. The House included a provision in the FY2013 defense authorization bill (H.R. 4310) providing $103 million to begin developing a plan and a supporting environmental impact statement to establish by the end of 2015 such a missile defense site on the East Coast to defend against a possible Iranian ICBM attack. That plan would evaluate the effectiveness, from an East Coast site, of various long-range BMD interceptor missiles.[3] The Senate version of the authorization bill does not include such a provision. The House FY2013 defense appropriations bill (H.R. 5856) did not include any such provision. Neither did the Senate defense appropriations bill.

Iran: A Source Global Political Concern

Iran is often a topic of discussion and a source of concern for many policymakers and experts in the national and international security community: "much of the debate over U.S. policy toward Iran has centered on the nature of the current regime; some believe that Iran ... is a threat to U.S. interests because hardliners in Iran's regime dominate and set a policy direction intended to challenge U.S. influence and allies in the region."[4] In his 2002 State of the Union address, President George W. Bush labeled Iran as part of the "Axis of Evil" along with Iraq and North Korea. More recently, the Obama Administration expressed its view of Iran as "a major threat to U.S. national security interests, a perception generated not only by Iran's nuclear programs but also by its military assistance to armed groups in Iraq and Afghanistan, to Hamas and other Palestinian militant groups, and to Lebanese Hezbollah."[5] To varying degrees these views are shared by many across the political spectrum and around the world. Yet there are considerable differences as to how to respond to Iran.

U.S. decision makers largely agree that Iran represents a threat to U.S. regional security interests. The 2010 National Security Strategy states that "for decades, the Islamic Republic of Iran has endangered the security of the region and the United States and failed to live up to its international responsibilities."[6] The United States "has long seen a threat to U.S. interests posed by Iran's support for militant groups" throughout the Middle East, Iran's support of the Syrian regime against growing popular opposition, and especially Iran's nuclear program, which "has emerged as a top priority for the Obama Administration."[7] Thus, Iran has been subject to a wide range of U.S. and international sanctions.

When coupled with its nuclear program, Iran's ballistic missile programs are a focus of major concern for a number of countries around the world. While discussing the implications of a February 2, 2009 Iranian satellite launch and addressing growing concerns over the threat posed by Iran, a Pentagon spokesman said such concerns affect Russia, Israel, the Middle East, Europe and beyond:

> We have long recognized that the threat—that they pose a real threat, and it is a growing threat, and that they are determined to develop long-range ballistic missiles. And I think all of Europe has recognized that to be the case, and that is why they unanimously embraced a third site for missile defense in Europe. And the Russians have recognized it, telling the secretary that they view Iran to be a threat as well. Clearly everybody in the Middle East recognizes it. The Israelis in particular but also the Gulf States are very concerned about Iran's actions in this realm. So this development today is cause for concern, not just here in the United States but in Europe, throughout the Middle East and, I believe, the greater world.[8]

Similarly, Russian Deputy Foreign Minister, Aleksandr Losyukov, was quoted by news agencies after a failed Iranian space launch attempt in 2008 saying the test "adds to general suspicions of Iran regarding its potential desire to build nuclear weapons" and he noted that long-range missiles "are one of the components of such weapons. That causes concern."[9]

Iran: A Source of Military Concern

More specifically, Iran's ballistic missiles challenge U.S. military capabilities and U.S. influence in the Middle East. According to the U.S. Intelligence Community (IC), "Iran already has the largest inventory of ballistic missiles in the Middle East, and it is expanding the scale, reach, and sophistication of its ballistic missile forces, many of which are inherently capable of carrying a nuclear payload."[10] According to the Pentagon, ballistic missiles in countries such as Iran threaten "U.S. forces, allies, and partners in regions where the United States deploys forces and maintains security relationships."[11]

Because of the size and capabilities of Iran's ballistic missile force, some argue that in the event of a crisis or conflict the United States might be deterred from engaging Iran militarily or that the possibility of ballistic missile attacks might preclude certain U.S. force options. In fact, the U.S. IC believes that Iran views its missile forces "as an integral part of its strategy to deter—and if necessary retaliate against—forces in the region, including U.S. forces. Its ballistic missiles are inherently capable of delivering WMD, and, if so armed, would fit into this strategy."[12]

In addition, some observers have argued that because many of Iran's neighbors lack their own deterrent capabilities or effective ballistic missile defenses Iran could "blackmail such states into meeting demands, for example, to raise oil prices, cut oil production or even withhold cooperation with the U.S. on which their very survival depends."[13] In the event of a conflict, one expert wrote, "U.S. military action would be to forestall Iranian interference with Gulf oil exports, this would have to be near total in its effect on Iranian capabilities. This would be difficult if not impossible to achieve, leading to a fear of attack which alone would have a formidable impact on oil prices."[14] Iran thus is viewed as a challenge to U.S. national security interests in the region, even with superior U.S. conventional military power. A major trend in U.S. military sales to the Persian Gulf states today is for anti-missile systems intended precisely to address this threat.[15]

Congress has long been concerned over the proliferation of WMD and their delivery systems, especially ballistic missiles. Over the past few decades, Iran has progressed from relying entirely on the outright purchase of ballistic missile systems to becoming nearly self-sufficient in important ways. Many now see Iran as a potential supplier to other nations or actors. This point is elaborated upon elsewhere in this report. Finally, some Members are concerned that an Iran armed with a nuclear ballistic missile, by the nature of the regime itself and the prospect of even wider proliferation that may follow, will result in a deterrence dynamic that is less stable and less reliable compared to the one that existed between the United States and the Soviet Union during the Cold War.

Finally, Iran's ballistic missile and space launch programs are major drivers of the U.S. BMD program. In addition to an extensive number of short-range ballistic missiles (SRBMs) and a smaller number of medium-range ballistic missiles (MRBMs) in Iran, the United States remains concerned about the possibility of an intercontinental range ballistic missile (ICBM) program in Iran. Since 1999, the IC has assessed that Iran could test an ICBM capable of striking parts of the United States by 2015 with sufficient foreign assistance.[16] Interestingly, however, LTG Burgess, Director, Defense Intelligence Agency (DIA) testified in 2010 that DIA "assesses that, with sufficient foreign assistance, Iran could develop and test an intercontinental ballistic missile (ICBM) capable of reaching the United States,"[17] but he did not include any particular date by which this might be accomplished. More recently, however, the 2012 DOD *Annual Report on Military Power of Iran* stated "Iran may be technically capable of flight-testing an intercontinental ballistic missile by 2015."[18] The report added that "beyond steady growth in its missile and rocket inventories, Iran has boosted the lethality and effectiveness of existing systems by improving accuracy and developing new submunition payloads,"[19] repeating what the same report said in 2010.

Potential Iranian missile threats to U.S. allies and friends in Europe and to the United States itself in the future are the basis for the Pentagon's plan to deploy increasingly capable phases of BMD, called the European Phased Adaptive Approach (EPAA), over the next decade in Europe and perhaps in the Persian Gulf as well.[20] At the May 2012 NATO Summit in Chicago, the Alliance declared an interim territorial missile defense capability based on U.S. ballistic missile defense (BMD) assets placed under NATO (North Atlantic Treaty Organization) command and control to guard against possible missile attacks from Iran and other potential adversaries.[21] The EPAA, which seeks to develop and deploy increasingly capable BMD over four phases through 2020, would serve as the cornerstone of this new NATO capability. Phase 1 of the U.S. EPAA was completed in December 2011 as scheduled. Phase 2 is expected to be completed by 2015, although Defense Secretary Panetta has stated that the future of the EPAA could be called into question by the application of automatic defense spending limitations under the Budget Control Act of 2012. The U.S. EPAA will be integrated with other BMD systems already being developed, augmented and deployed by NATO individual member states under NATO command and control architecture.[22]

Further, some analysts, including the 1998 Rumsfeld Commission on the evolving missile threat to the United States and its allies, suggest that countries could threaten U.S. territory by pursuing unconventional deployment modes.[23] Among these, it is argued, is to deploy shorter range missiles on surface ships. These analysts argue that such ships could be brought close to coastal areas of the United States, thus giving the shorter range missiles the ability to strike targets in the United States. Others discount the ease with which this could be accomplished.

CONTEXT

This report is limited to analyzing Iran's efforts to develop, test and field ballistic missiles and to develop a space launch capability. Iran has pursued these efforts with the assistance of other countries and entities operating in those countries with and without direct or indirect official government sanction. Assessing Iran's progress and capabilities in this area are challenging given the general difficulty in evaluating ballistic missile capabilities and projections in general, especially from the vagueness of official assessments and the numerous, but specific and conflicting analyses by experts and others in the public domain. Although Iran is transparent in many ways, it remains secretive and attempts to obfuscate intent and capabilities in others. The latter tendency frequently extends to matters related to advancing its military technologies. All these things contribute significantly to the difficulty inherent in such a report as this. These issues are addressed below.

Scope

In and of itself, the acquisition of ballistic missiles is not necessarily problematic to U.S. interests. Indeed, many U.S. friends and allies have such missiles and pursue missile modernization programs. But when a ballistic missile program is pursued by a state hostile to the national security interests of the United States and its friends and allies, such an effort becomes a source of concern. Such concerns are amplified considerably when that state might develop a weapon of mass destruction (WMD), particularly a nuclear weapon that could be carried by that ballistic missile force. In Iran's case, no major intelligence community has concluded that Iran has made the decision to develop a nuclear weapon. But Iran has taken steps in its civilian nuclear program that raise serious questions about future intent to develop or break out of its nonproliferation commitments and obtain nuclear weapons on short notice.

An effective nuclear weapons capability has three major elements, each of which presents its own unique technical and other challenges. Additionally, each of these elements must work together as part of an operational system. All this is required to make the case that a specific ballistic missile threat is grave and imminent.

First is the production of fissile or nuclear material in sufficient quantity and quality as the source of a nuclear device.[24] The second element concerns the design, weaponization, miniaturization, and survivability of the warhead and its physics package capable of being delivered to its intended target, especially at long-range.[25] The third element is providing an effective means for delivery, such as a ballistic missile or, as some believe, a capability that could grow out of a space launch program.

This report focuses solely on Iran's ballistic missile and space programs. It is important to note that this report does *not* assess Iran's capabilities in the other two critical areas. Without a reliable, survivable nuclear warhead the proliferation of Iranian ballistic missiles is arguably not an imminent significant threat.

Finally, there are literally many thousands of articles, news accounts and other sources regarding Iran's ballistic missile and space programs. It is not possible to sort through the tremendously varied and often different accounts of events or capabilities. This report does not rely on those items or try to adjudicate any differences in them. As much as possible, this report instead emphasizes U.S. Government and other official sources. It is noted, however, that official sources are frustrating to rely on as they frequently do not provide much specificity. On the other hand, some of the best non-government open sources provide much more specificity yet those same sources can vary considerably from one to another on the very same elements of Iran's ballistic missile and space launch programs. There are, however, perhaps a dozen or so excellent analysts who write and publish on Iran's programs, but who may differ on their various facets. This report attempts to make clear where some of those key differences exist.

The Technological Challenge of Ballistic Missiles

This section very briefly touches on some of the challenges inherent in rocket science and missile engineering regarding the development, testing and production of ballistic missiles.

As enormously capable as the United States is regarding the development and acquisition of ballistic missiles, it too experiences challenges and even failures on occasion. For instance, several years ago when the Pentagon sought to develop the first variant of the Ground-based Interceptor for long-range BMD, it experienced significant technical problems in developing a new solid rocket motor. Eventually, the Pentagon chose to go forward with an interceptor based on existing launch vehicles (Minotaur and Pegasus). Other modern states with advanced missile and space launch programs will also run into problems while developing new missile systems despite their long-standing expertise and an extensive industrial base that is unencumbered with external trade constraints, such as sanctions or export controls. In contrast, Iran's expertise and industrial base is significantly challenged in this regard on a number of fronts and cannot be viewed as comparable to that of the United States or of the other major nuclear powers. Nevertheless, Iran has demonstrated persistence over decades in continuing to develop and improve its indigenous capabilities in spite of widespread international opposition and cooperation against Iran's ballistic missile programs. A significant advantage that Iran has had is that the basic technology of rocket science is established and reasonably well known around the world.

Another comparative marker is to contrast the number of missile flight tests between modern missile countries such as the United States and Russia on one hand and Iran on the other. By the end of their development efforts, the United States had conducted perhaps a couple dozen or more flight tests of the Trident II submarine-launched ballistic missile and somewhat less than that number for the MX ICBM. The Soviet Union/Russian flight test record for their ICBMs (SS-24, SS-25, and SS-26) was comparable. Moreover these flight tests took advantage of decades of flight testing of older systems dating to the 1950s. This record entailed scores of space launch flight tests and over a hundred each of the earliest ICBMs (Atlas and Minuteman I) prior to their deployments. In contrast, Iran has conducted very few flight tests of its modern missiles. This has major implications for how much confidence Iranian decision makers and military leaders would have in the reliability and expected performance of their ballistic missile force to achieve its intended missions. Iran may also have difficulties in developing rocket motors for various reasons some of which might include the effect of sanctions and export control regimes. The U.S. IC has touched on this on many occasions. Although many principles of rocket science are much better known and more widely shared today, missile engineering remains a challenging skill acquired only through direct experience—it cannot simply be studied and applied without much trial and error. For instance, there is tremendous complexity in finding the right formulation of solid propellants for a specific missile system. Although the underlying rocket science has been reasonably well understood and publicly available for decades the predictability of how a solid rocket motor will perform depends on very subtle factors. This means that even after a propellant is formulated, the manufacturing of solid rocket motors still requires extreme reliability controls.[26] Additional factors necessary for developing an effective ballistic missile include a guidance system that directs the missile and its payload to its intended target. This requires testing of missiles and their reentry vehicles, which produces data from which mathematical models can be developed, further tested and fine-tuned. From the time a ballistic missile lies dormant in its silo or launch vehicle to its actual launch, various errors can begin to accumulate. A systematic and rigorous testing regimen of all the component parts can help reduce as much as possible such structural or inherent errors even in proven systems, but such errors cannot be eliminated entirely. Operational tests of the system are generally necessary to know whether the entire system will work as intended, and the degree to which further testing may be required to ensure such missiles will launch successfully and operate as planned.

Management and Organization

Many experts would argue that strong management and organization are considered critical to a good ballistic missile program. Iran is generally assessed to have such an organizational and managerial program, although it too cannot be compared to similar programs in the major nuclear weapons states—the United States, Russia, China, the UK and France. Michael Elleman, for instance, among others, has said "Iran's record in the area of overall systems management and engineering involving a range of technologies and industries is impressive."[27] He has further said

> their approach to missile development and modification is disciplined and well structured. This is unique in the proliferation world. They are not up to Western standards, but they are far more sophisticated than the DPRK, Iraq or even Pakistan. This capability empowers Iran to develop and produce whatever missile range it decides the country needs. It may take a decade or more, and they will still rely on foreign sources for some key items, but there appears to be little the West can do to stop (as opposed to impede) progress.[28]

Aaron Karp, one of the first experts to write about this, said

> There is much more to any major R&D project than just assembling metal and plastic. Easily overlooked are the necessary skills, experience and judgment required of engineers and programme managers. Also, behind every missile programme are conceptual, organizational, financial, and command and control factors, each imposing its own problems for ballistic missile development.[29]

Karp also said that, although countries may receive considerable amounts of foreign assistance in their missile programs, "would-be rocket makers are almost entirely on their own" in the area of program management: "foreign companies and governments can offer advice and their own example, but little else." For instance, both appropriate organization and managerial choices have proven critical to successful ballistic missile programs when governments are choosing their development strategies.

An incremental development strategy, which Iran appears to have pursued, in which a missile program moves in sequence through progressively larger designs, improving the performance of major components, and gradually introducing new ones is the most effective according to Karp. He cites the experience of the French as the best example of this. By contrast, states that have begun their missile programs by taking huge leaps tend to face greater difficulties. For instance, the UK began by developing an IRBM using this approach and that program eventually collapsed. Similarly, India began its space launch program in the early 1970s enduring many false starts and serious delays.

Missile Proliferation: Challenges with Estimates and Projections

Assessing ballistic missile proliferation capabilities in general has proven to be challenging. Further, projecting those capabilities over some near- or long-term future has proven unreliable.

As a result, efforts to time responsive measures precisely according to program advancements in select countries are likely to fail by projecting such advancements coming later or earlier than will turn out to be the case. In the first instance, the United States was surprised by the advancement of the Soviet Union's space rocket capability, and by implication its ballistic missile program, with the launch of the Sputnik satellite. At the other end of the spectrum, in 1990 the U.S. IC projection for MRBMs and ICBMs by 2000 did not transpire:

> Most missiles likely to be fielded in the Third World over the next 5 years will have ranges of less than 1,000 kilometers, but by the year 2000 at least six countries probably will have missiles with range of up to 3,000 kilometers. At least three of them may develop missiles with ranges of up to 5,500 kilometers.[30]

Instead, intelligence assessments in 2000 showed that no country in the developing world had acquired such capabilities.[31] And the 1990 forecast has not been borne out through today.

For a variety of reasons, projections of ballistic missile trends in countries over time may not materialize as expected, instead slowing significantly, apparently halting, or even reversing from earlier assessments.[32] In some cases, missile proliferation estimates might show significant projected increases from year to year only later to show that same effort apparently halting entirely.[33]

In particular, few countries have successfully developed and deployed operational, nuclear-armed ICBMs. The developmental record of those who were successful indicates how challenging that effort was.[34] The fact that more nations have not achieved this capability is perhaps witness in part to the extraordinary technical effort it took. The long history of acquiring ICBM capability among the five nuclear-armed ICBM states (U.S., Russia, China, UK[35] and France) demonstrates that such success required considerable resources in time, funding, knowledge, infrastructure, organization and national commitment. After 50 years of commitment and experience, these nations still experience developmental or even operational failures in their missile or space launch programs on occasion. Although there are frequent charges of the ease in which a country might develop an intercontinental ballistic missile from a space launch program, this has not in fact occurred under missile programs by emerging missile and rocket powers, despite the fact that so many of the technologies are comparable. Perhaps the closest example is that of India, which may have used a few elements of its more than 30-year old space program to develop and test an ICBM-range missile for the first time in April 2012.[36]

Some argue, however, that the small number of ICBM programs to date may result from the small number of countries who perceive the need for ICBMs rather than it being so difficult. Five countries make ICBMs, but nine countries have made successful SLVs, which are close to ICBMs. Additionally, it is argued, that countries such as the United States, the Soviet Union and France took considerable time and money to build their ICBMs is not necessarily an indication for other ICBM contenders. The superpowers and France adhere to high quality and reliability standards, which may or may not be present in Iran. Iran also may be satisfied with an ICBM capability of lesser quality and reliability, it is argued.

Contributing further to understanding missile proliferation in Iran is the difference in names often given to various Iranian missiles by Iran, the United States, U.S. friends and allies, and other good open sources, even when it appears specific missile capabilities seem to suggest the same missile is being described. This report does not attempt to reconcile those differences, but rather retains the missile designations in the sources cited.

Iran: A Special Case in Opacity

In many respects, a strong argument can be made that Iran is somewhat transparent in its ballistic missile activities. Iran regularly televises and talks about its operational missile exercises, publicly showcases many of its ballistic missile and space launch tests, and even holds televised interviews with Iranian officials and leaders in some of its missile and nuclear plants. By comparison, some countries considered allies or friends of the United States do not demonstrate this much openness themselves. In the absence of a diplomatic relationship or any kind of verifiable arms control regime between the countries, a few observers have suggested that Iran may be signaling a certain level of transparency to the United States and others regarding its ballistic missile and space launch programs.

Iran readily acknowledges that U.S. and other countries "possess satellite systems, especially the countries which control the space above Iran, [who] can see our missiles quite clearly. That is, their satellite systems, and not their radars, pick up the movement of our radars, including the firing platform and even the landing site. It is quite clear to them" when maneuvers are real and when Iran exercises with live weapons and are not simply exhibitions.[37] After two MRBM tests in February 2011 that Iran said were observed by the U.S. military, Iran expressed surprise that the United States chose not to disclose those launches publicly. Uzi Rubin, an Israeli missile engineer, former head of Israel's BMD program, and who is also a critic of Iran's ballistic missile programs, said the Iranians are transparent in these types of tests, openly wishing for "U.S. satellites to take pictures of their weapons sites and to see their capability."[38]

On the other hand, Iran has proven itself capable of misleading and deceitful statements regarding its ballistic missile and space launch programs. It is useful to briefly point this out as it casts serious doubt on relying heavily on Iranian statements about their ballistic missile and space launch programs. Although some U.S. agencies do that, this report takes exception to that practice. Although this report sources many Iranian statements about their missile programs and tests, it does not assume those statements are accurate without further corroboration or notes that Iranian claims are not confirmed. Although some examples of Iranian misinformation are scattered throughout this report, some of the more notable deceits include

- a photo-shopped "successful" test of a failed Iranian SRBM launch in 2008;[39]
- a "successful" launch of a "long-range radar-evading" ballistic missile in 2006,[40] which was actually video of a missile launch from a Chinese submarine;[41] and
- the "successful" launch of an Iranian satellite into orbit in 2008,[42] which actually exploded shortly after launch.[43]

It is possible that the Iranian media, rather than Iranian officials account for some of the distortions reported. Some have also raised questions over the accuracy of translations from the Persian language, which could then be read after translation as deliberate misinformation or misleading reporting from official sources.

IRAN'S BALLISTIC MISSILE AND SPACE LAUNCH PROGRAMS

Iran has been acquiring, developing and testing its ballistic missile capabilities for decades. This section examines the long-term investment Iran has made in developing ballistic missiles and in building an extensive network of facilities. It further examines Iran's interest in and capabilities for a space launch capacity as well. What is striking is the degree to which there is an overwhelming amount of information written about Iran's ballistic missile capabilities, performance parameters, testing results, infrastructure and Iran's cooperation with others compared to the paucity of public detail about how many missiles Iran actually has.

Context and Historical Background

Iran's interest in developing ballistic missiles can be traced back to the 1960s and can be tied to its regional security interests even before the Islamic Revolution in 1979. In general, the U.S. intelligence community has publicly stated that it believes Iran's overall approach to its international affairs "will remain relatively constant and will continue to be driven by longstanding priorities of preserving the Islamic regime, safeguarding Iran's sovereignty, defending its nuclear ambitions, and expanding its influence in the region and the Islamic world."[44] Similarly, the Defense Intelligence Agency (DIA) stated the strategic objectives of Iran's leadership are "first and foremost, regime survival; making Iran the preeminent regional power; and turning Iran into an economic, scientific, and technological powerhouse."[45]

Currently, the U.S. intelligence community believes Iran's growing inventory of ballistic missiles, as well as its production of anti-ship cruise missiles, provide capabilities to enhance its power projection. Dennis Blair, then Director of National Intelligence, said in his 2010 annual threat assessment report to Congress that

> Iran already has the largest inventory of ballistic missiles in the Middle East and it continues to expand the scale, reach and sophistication of its ballistic missile forces – many of which are inherently capable of carrying a nuclear payload....Tehran views its conventionally armed missiles as an integral part of its strategy to deter – and if necessary retaliate against – forces in the region. Its ballistic missiles are inherently capable of delivering WMD (Weapons of Mass Destruction), and if so armed, would fit into this same strategy.[46]

Similarly, the DIA said in 2010 that Iran

> continues to develop ballistic missiles capable of targeting Arab adversaries, Israel, and central Europe, including Iranian claims of an extended-range variant of the Shahab-3 and a 2,000-km medium range ballistic missile (MRBM), the Ashura. Beyond the steady growth in its missile and rocket inventories, Iran has boosted the lethality and effectiveness of existing systems with accuracy improvements and sub-munition payloads.[47]

There is no reliable open-source assessment as to how much Iran spends to develop, test, and field its ballistic missile capabilities, nor how much it spends relative to other nations that develop, test, and field ballistic missiles. There is limited open-source data, however, regarding Iran's military expenditures in general. Over the period 1997-2009, Iran reportedly spent about $6.2 billion annually (U.S. constant dollars).[48] Iran's military expenditures placed it fourth among the other Gulf States and the GCC.[49] As a function of GDP, Iran's military expenditures placed it sixth among those same states[50] and 60th in the world according to the CIA.[51]

Iran's Ballistic Missile and Space Launch Complex

Unclassified government and other open-source materials indicate that Iran has an extensive ballistic missile and space launch complex scattered throughout the country.[52] This section highlights only some of those reports. Figure 1, "Iran's Ballistic Missile and Space Launch Complex," illustrates where some of these facilities are apparently located. These include various short- and medium-range missile sites, missile test and space launch sites, and a number of sites suspected to be involved in the production, assembly and storage of ballistic missiles and missile-related infrastructure.

Some of Iran's ballistic missiles have reportedly been deployed at some time or another near Tehran, Tabriz, Kermanshah, Khorramabad, Semnan, Sharud, Shiraz and Mashad. Iran is known to have flight tested ballistic missiles of varying ranges and at different times in many locations; some of these locations may lie dormant, while others are more active. They include sites near Qom, Qeshm Island, Garmshar, Sharud, Semnan, and Tabas for instance. Although Iran currently relies primarily on the Semnan Space Launch complex in north central Iran,[53] it also has reportedly conducted some space launch activities near Qom and perhaps elsewhere.[54]

The Islamic Revolutionary Guard Corps (IRGC) is part of the duly constituted armed forces, but its primary role is to protect the Islamic regime.[55] It is the IRGC Air Force that operates Iran's ballistic missile forces.

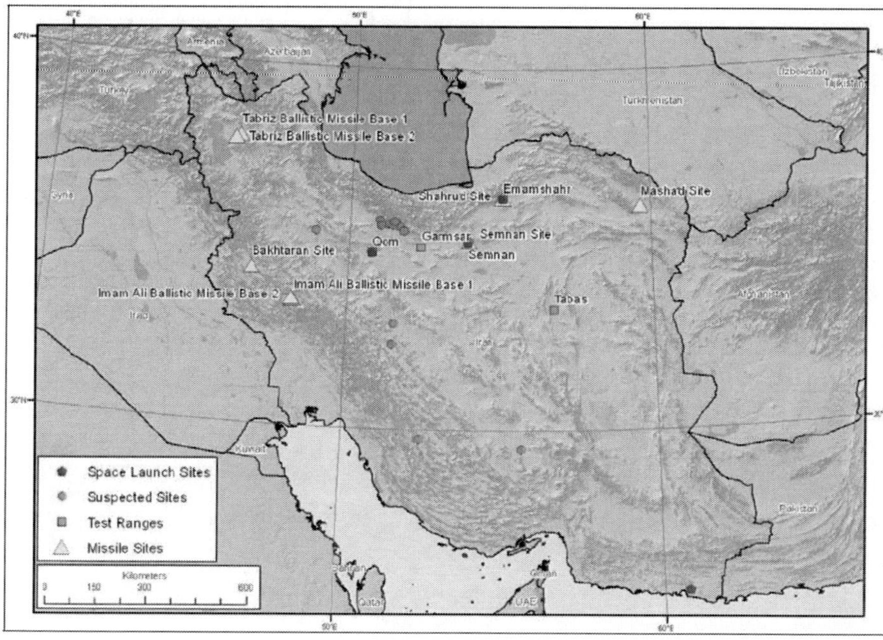

Figure 1. Iran's Ballistic Missile and Space Launch Complex.

Short-Range Ballistic Missiles (SRBMs)

Short-range ballistic missiles (SRBMs) are generally defined as those ballistic missiles with ranges of less than 1,000 km. Some analysts further differentiate tactical ballistic missiles as those with ranges between 10-300 km. Iran views its SRBMs as an important part of its conventional military capability. The DIA said in 2010 that SRBMs provide Iran "with an effective mobile capability to strike coalition forces in the region. Iran continues to improve the survivability of these systems through technological advances, such as solid-propellant and the use of anti-missile defense tactics."[56]

A precise, public accounting of Iran's SRBM force is not available. Official U.S. sources often cite the figure of "hundreds of SRBMs" with perhaps 50-100 launchers (a launcher can be reused to fire additional missiles). There is no further breakdown of these numbers. See Table 1, "Iranian SRBMs (U.S. Government)." Table 1 does not include an estimate for some years as nothing was noted in that particular year. Figure 2, "SRBM Sites and Ranges," illustrates the locations and the estimated operational SRBM ranges from those sites. The dotted ring around Shiraz roughly illustrates the operational range at a typical SRBM site.

Other good public sources cite figures of perhaps 200-300 Shahab-1 and Shahab-2 SRBMs (with as few as 18-20 launchers or up to around 50 launchers). See Table C-1, “Iranian SRBMs (Other Sources).” These latter sources cite additional, different named SRBMs in Iran’s inventory, which this report does not attempt to deconflict. This section provides a brief overview of Iran’s SRBM missiles and programs.

Finally, imagery obtained through the Library of Congress shows what appear to be a number of short-range missile bunkers in western Iran around Kermanshah and what a SRBM missile base might look like. See Figure 3, “Kermanshah SRBM Base.”

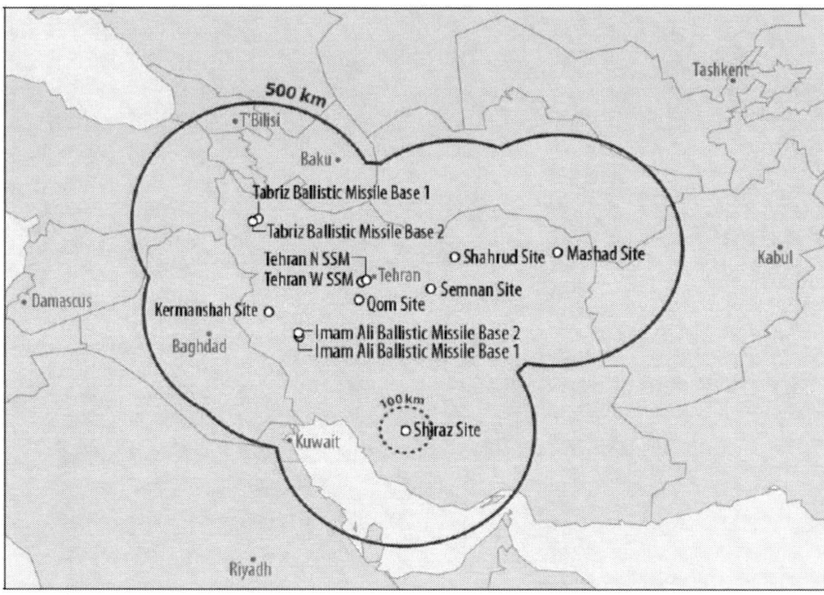

Figure 2. SRBM Sites and Ranges.

Source: CRS.

Figure 3. Kermanshah SRBM Base.

Shahab 1 and Shahab-2

Iran's initial, limited purchases of SRBMs came from Libya and Syria during the Iran-Iraq War (1980-1988). These missiles were used to strike targets deeper in Iraq from points further removed from the war's battlefront.[57] Later in the war, Iran acquired large numbers of Soviet Scud B and Scud C SRBMs primarily from North Korea. Iran began calling these missiles Shahab-1 and Shahab-2.

Both are liquid-fueled ballistic missiles now produced by Iran, which developed an indigenous infrastructure with the help of the DPRK after the Iran-Iraq War to assemble these missiles. Iran likely continues to rely on some foreign-produced components to produce the Shahab-1 and Shahab-2 missiles:[58] "Uncertainty remains over Iran's ability to produce the Shahab-1 indigenously," and although it can "reliably produce the Shahab-2 missile," Iran continues to rely on foreign sources.[59]

Both the Shahab-1 and the Shahab-2 are road-mobile systems that can be moved to any number of pre-surveyed launch sites. Wartime experiences, such as in Iraq, show these missiles tend to operate within a radius of about 100 km or less from their bases because of the need to ensure operational security and to be able to maintain key logistics support. A notional operating area of these missiles is shown in Figure 2 around Shiraz in southwestern Iran.

The Shahab-1 has a range estimated of about 300 km, while the Shahab-2, which carries a lighter conventional warhead, has a range estimated at about 500 km. Iran has an unspecified number of Shahab-1 and Shahab-2 SRBMs, but they likely number on the order of a few hundred.

Qiam

The only reported test of this ballistic missile in the media was in August 2010. A UN Report said a Member State "assessed the Qiam to be based on the Shahab-2, with a range between 500 and 1,000 kilometers. Some experts have raised questions about the missile's lack of apparent testing. Missiles are known to require extensive flight testing before they can be fully operational."[60]

The status of the Qiam is unclear. According to a Hezbollah (Lebanese) media source, the Qiam 1 was delivered to the Aerospace Force of the IRGC in late May 2010.[61] Iran's Islamic Republic News Agency quoted Defense Minister BG Vahdi as saying the Qiam is harder to detect than older models and that mass production of the Qiam missile, the country's first missile without stabilizer fins, demonstrates Iran's self-sufficiency in producing various types of missiles. Gen.

Vahdi added that the Qiam's design reduces the possibility of being detected by enemy anti-missile systems and the omission of its fins increases the missile's speed enabling it to hit its targets with high precision. According to the same Hezbollah source, IRGC Commander Mohammed Ali Jafari, told reporters earlier in 2010 "these new missiles enjoy supersonic speed and cannot be tracked or intercepted by the enemy."

Before Syria's civil war, some in Israel thought the Qiam might be exported to Iranian clients, such as Hezbollah in Syria.[62] This appears less likely today, however, given conditions within Syria.

Table 1. Iranian SRBMs (U.S. Government)

	2000	2001	2002	2003	2004	2005	2006	2007	2008	2009	2010	2011	2012
U.S. Intelligence Community Assessments*													
CSS-8 Scud B Scud C	Several hundred SRBMs		Several hundred SRBMs		A few hundred SRBMs	A few hundred SRBMs		Hundreds of SRBMs	Hundreds of SRBMs				
Fateh-110				In production									
National Air and Space Intelligence Center (U.S. Air Force), Reports on Ballistic and Cruise Missile Threats**													
CSS-8	< 50 launchers			< 50 launchers			< 50 launchers			< 100 launchers			
SCUD B	< 50 launchers			< 50 launchers			< 50 launchers						
SCUD C	< 50 launchers			< 50 launchers			< 50 launchers						
Fateh-110				not yet deployed			< 50 launchers						

Sources: *Annual Threat Assessments, 721 Reports. **NASIC (U.S. Air Force), Reports on Ballistic and Cruise Missile Threat. Missile designations as used by the source.

Notes: Where launcher numbers only are noted, actual missile inventory may be larger because launchers can be reused to fire additional missiles.

Fateh-110

The Fateh-110 is a solid-fuel, road-mobile battlefield or tactical ballistic missile with a range of about 200 km. Its development probably started around 1995 and its first test flight reported in 2001. There may be three versions of the Fateh-110 in service; one is apparently called the Khalij-Fasr.

An upgraded version of the Fateh-110 reportedly was tested in early August 2012. Iran claimed it was the fourth generation of the Fateh, and equipped with a new guidance system capable of striking targets up to 300 km away with high accuracy.[63] Experts such as Elleman find the Iranian assertion dubious; the missile has a maximum range of only 200-250 km, and claims of high accuracy are questionable.

Medium-Range Ballistic Missiles (MRBMs)

Iran has invested for several decades in an organized and well-planned effort to acquire, develop, test and deploy MRBMs capable of striking targets throughout the Middle East. Iran views these missiles as a deterrent to any military action that might be taken against it and has said that it can now target about three dozen U.S. bases throughout the region. Some number of these missiles may be housed or deployed in underground bunkers or silo-like structures. The United States and others have long maintained that Iranian MRBMs are inherently capable of carrying nuclear warheads.

No official numbers of MRBMs operationally deployed are provided by official U.S. government sources. Those numbers that do exist show growth from less than 20 MRBM launchers to less than 50 MRBM launchers by 2009. Actual numbers of the missiles are not publicly provided by the U.S. government. Table 2, "Iranian MRBMs (U.S. Government)," shows that neither the annual U.S. threat assessment nor the annual 721 proliferation reports[64] provide any numbers of MRBMs or launchers. Only the U.S. Air Force has provided an official, public number of launchers, albeit not detailed. Table C-2, "Iranian MRBMs (Other Sources)," provides additional information, but it too illustrates the lack of preciseness as to how many MRBMs Iran has deployed. Other countries, such as Israel, have reported large numbers of MRBMs (i.e., 450) that could not verified by others.[65]

Figure 4, "MRBM Sites and Ranges," illustrates possible ranges of Iranian MRBMs from two apparent missile silo launch complexes in western Iran and from Iran's ballistic missile and space launch complex in central Iran. This gives one an idea of the area potentially covered by Iranian MRBMs.

In 2010, the DIA provided about as much specificity as has been seen about Iranian MRBMs, stating that Iran has "developed medium-range ballistic missiles, and continues to increase the range, lethality, and accuracy of these systems. The Shahab 3, based on the North Korean No Dong, can reach all of Israel. The Ashura or "Sejil" is an indigenous, two-stage missile under development, which uses solid-propellant technology, reducing the launch preparation and footprint."[66] Iran faces geographical constraints in testing its ballistic and space launch systems. In a lengthy 2006 interview about Iran's ballistic missile programs, then IRGC Commander Gen. RahimSafavi made this point. He said, "the longest distance from the northwestern region of Maku to Chabahar is about 2,250 km. We usually fire towards Iran's central desert [and] because Iran is shaped like a trapezoid, we can fire along the longest diameter of the trapezoid. We also watch very carefully to make certain that there are no villages or other inhabited areas along the firing range of these missiles, lest they should be hit."[67]

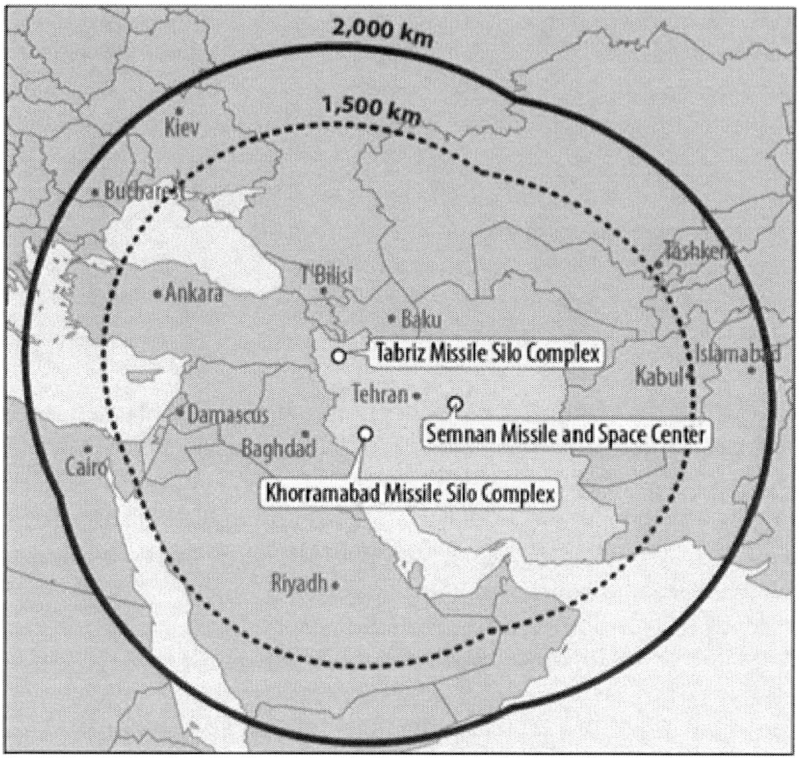

Source: CRS.

Figure 4. MRBM Sites and Ranges.

Therefore, estimates of some of Iran's MRBM tests may be based on observing missiles fired non-optimally, such as in a lofted trajectory that sends missiles higher than they would be relative to an optimal ballistic trajectory. Or these missiles might be flight-tested on nominal or minimum energy trajectories. There is no public information available on this.

It is unclear if Iran possesses adequate telemetry gathering capabilities needed to determine flight test results in a true ballistic trajectory over the ocean. On one hand, if Iran's MRBM tests have been lofted or on nominal trajectories, landing over land, this might suggest they lack that capability. Additionally, various technical assessments, such as the UN Panel of Experts reports cited earlier, indicate Iran lacks many things in its ballistic missile programs, including telemetry and instrumentation devices.

Uzi Rubin, however, claims that Iran's recent missile tests landed in the mouth of the Indian Ocean and thus argues that Iran has some capability to deploy instrumented ships or aircraft near the missile's impact points. Iran would do this "to obtain data on the behavior of the missile toward the end of its flight" and that such instrumented platforms could be deployed increasingly down range in the Indian Ocean "to test ballistic missiles to any range it wants."[68] Rubin believes the speed of Iranian recovery from failures is such that it implies plentiful telemetry data. Further, he argues, a successful space program would necessitate a full suite of telemetry and ground support equipment. This section provides a brief overview of Iran's MRBMs.

The Shahab 3 and Its Variants

The Shahab-3 is a ballistic missile imported from the DPRK and based on the No-dong 1. Iran may have "purchased a production line for these missiles in the early 1990s and is now manufacturing them rapidly," according to one expert.[69] The Shahab-3 has been given various names by Iran and others over time. There reportedly have been several different versions of this liquid-fueled missile flight tested with various other modifications made to it, perhaps providing the Shahab-3 with ranges varying from about 800 – 1,000 km.

Modified Shahab-3/Ghadr-1/Kadr

Because the range of the Shahab-3 lies at the low end of an MRBM, Iran has sought to develop, test and deploy a much longer range ballistic missile.[70] It has sought to do this with both liquid-fueled missile programs based off the Shahab-3, such as the Ghadr-1 or Kadr (or Kadr F), and a separate solid-fuel missile such as the Sejil or Ashura.

Iran first pursued a number of efforts to extend the range of the Shahab-3 with mixed test results. One expert wrote that "there are indications that perhaps as many as one-half of them [Shahab-3 tests] failed."[71] Some have variously referred to this as an extended range or modified Shahab-3 or Ghadar-1. Range estimates for this missile vary widely, from 1,000 km to 2,000 km. Such range differences could in large part be explained in terms of ranges associated with payloads of different weights.

This missile reportedly carries a smaller payload than the Shahab-3, includes a lighter airframe for greater distances, and has "an improved guidance system and uses a triconic aeroshell geometry that provides greater aerodynamic stability."[72] The Ghadar-1 design, "almost certainly includes a separating warhead," could provide the capacity "to carry airburst warheads, or warheads containing submunition packages."[73] It was reported in 2006 that some Western intelligence experts believed Iran had been able to modify the nose cone or reentry vehicle (RV) of the Shahab-3 with Russian and Chinese assistance. An identical RV was twice displayed on No Dong missiles in North Korean parades, so it may have been designed by North Korea. One source says that "instead of the single cone normally attached to this type of missile, the new Shahab has three cones, or a triconic, warhead. A triconic warhead allows the missile to accommodate a nuclear device and this type of warhead is normally found only in nuclear weapons."[74] But others have suggested the triconic RV may be less suitable for a nuclear payload than the previous conical RV associated with the older Shahab 3, and may instead be designed for higher accuracy which may indicate a nonnuclear mission. But none of this has been officially confirmed in Iran. U.S. and other intelligence reports indicate that Iran apparently has not made the decision to develop a nuclear weapon, but developments such as these cause concern among many decision makers and observers.

What of the Shahab-4, Shahab-5 and Shahab-6?

Various reports of a Shahab-4 surfaced in the 1990s and those missile designations are still used or mentioned on occasion. It appears, however, that whatever was intended originally, any "Shahab-4"missile effort likely led to the development of the extended-range Shahab-3 MRBMs or the Ghadr-1.

In the late 1990s and early 2000s, other reports said Iran was developing a solid- or liquid-fueled, two- or three-stage MRBM or space launch vehicle. It is possible they once had a designation of Shahab-5 and Shahab-6, which continues to cause some confusion today. But, more likely these programs led to other ballistic missile and space launch programs in Iran today. One may have become or led to the Ashura/Sejil and the other may have become or led to the Safir space launch vehicle.

When asked in a lengthy and seemingly rather open interview in 2006 about Iranian military exercises ('Great Prophet' IV) and Iranian ballistic missiles, then IRGC Commander Gen. Safavi said, "I'd prefer not to answer," whether there were Shahab-4 and Shahab-5 missiles.[75]

Sejil/Sejjil 2/Ashura

The Sejil is a solid-fueled ground-mobile ballistic missile that Iran says has a range about 2,000 km according to official U.S. sources. Both Postol and Elleman state a 2,200 km range. Some in Israel and elsewhere believe the missile might have a range up to 2,500 km.

The Ashura was announced by Iran in November 2007. This announcement was accompanied either by an unsuccessful test of the solid propellant motors or the Ashura missile itself depending on whether an Israeli or Russian account, respectively, was accurate.

Subsequently renamed the Sejil, it was flight tested successfully in November 2008, reportedly at about 800 km. Additional tests of the Sejil or a modified and apparently more capable version of it (Sejil 2) took place in May, September and December 2009. From the progress shown in the Sejil solid-fuel rocket program, Elleman concludes that "Iran is at a minimum in the process of mastering the technology."[76] Many consider the Sejil 2 a nuclear capable ballistic missile as its payload capacity could accommodate a first generation nuclear warhead. Some have suggested that the MRBM silos near Tabriz and Khorramabad might be intended for the Sejil 2. Many believe the missile is the same regardless of the naming designation (i.e., Sejil, Sejjil 2, or Ashura).

The R-27/BM-25/Musadan Controversy

The issue of whether Iran received at some point some number of Soviet-era R-27 SLBMs (submarine launched ballistic missiles) or components remains unsettled, controversial and persistent. Early on, the German press named this missile the BM-25 when they referred to the Iranian version, and others have referred to this as the Musadan when used in the context of North Korean involvement. The pedigree of this system is far from trivial and its reported design or redesign remains obscure. No official public U.S. assessment or word of this was found.

This issue, therefore, has been largely been a debate among technical experts and observers who track Iranian missile programs. Whether Iran acquired these missiles or their rocket motors in whole or part, or received other technical assistance regarding these missiles or their component parts is important. If accurate, acquiring this capability or technical knowledge might enable Iran to build more powerful and longer range ballistic missiles.

Some experts lend credence to assessments that Iran may have some number of these missiles or their components. Uzi Rubin stated that "Iran has acquired eighteen BM25 land-mobile missiles with launchers from North Korea, which can strike targets in Europe. In the past, the BM-25 has been produced in two models: one with a range of 2,500 km and the second with a range of 3,500 km."[77] Similarly, according to an unpublished May 2011 UN

Panel of Experts Report, two Member States shared the assessment that "Iran received a shipment of 19 BM-25 missiles from the DPRK in semi-knock down and complete knock down kits.[78] German missile scientists Schmucker and Schiller acknowledge that the BM-25 is "seen as an operational part of the North Korean and Iranian missile arsenals, though it was never actually launched," but that its display in a 2007 DPRK military parade turned out to be a different, very short-range missile.[79]

The *Washington Post* reported in 2010,[80] citing U.S. cables, that Iran had obtained 19 of the BM-25 or Musadan missiles from North Korea. The document reportedly summarized a meeting of U.S. and Russian technical experts and officials, where the Russians claim the BM-25 might not even exist and U.S. officials acknowledge never seeing the missile in Iran. It turns out that the U.S. delegation reportedly relied heavily on a 2005 article from a conservative leaning German tabloid called *Bild Zeitung* quoting German intelligence sources that Iran had purchased 18 (not 19) BM-25 kits made up of missile components for the BM-25 from North Korea. One technical expert was quoted in the *Post* article as saying "the U.S. side does not firmly say we have evidence that the BM-25 is in Iran."

This missile has not been seen publicly in Iran and has not been tested. Some were reportedly shown in North Korea, but subsequently discounted as mock-ups. Some analysts have reported the BM-25 might be capable of reaching 3,500 km. It is reputedly based on a Soviet-era submarine launched ballistic missile known as the R-27 in Russia and the SS-N-6 in the West.

Experts such as Michael Elleman believe that although it is "highly improbable that complete or disassembled R-27 missiles were exported by Russia, it is possible that individual components of the missile may have been smuggled out of the country."[81] Other missile experts such as Ted Postol of MIT argue that the evidence suggests that an unknown quantity of Soviet-era naval ballistic missile parts were shipped to North Korea without the approval or knowledge of the central government during the collapse of the 1990s. Russia never acknowledged this transfer, Postol and Elleman said, because it would tarnish Russia's reputation as a country that claims to have never sold technology that could be used in an intercontinental ballistic missile.[82] Over time, it appears that some of these missile parts and components, or at least their design features, seem to have made their way into North Korean and Iranian missile and space launch programs. Postol has made this argument in assessing the Iranian Safir space launch missile, which has a second stage that appears to use the two vernier thrust chambers and common turbopump from the R-27. In addition, the unusual spatial arrangement of the two thrust chambers, their hydraulic mounts and the conical shape of the stage's back-end are essentially the same as that on the R-27.[83] Some of this is touched on briefly in the section on Iran's space launch program.

MRBM Silo-like Launch Facilities

During the June 2011 "Great Prophet" military exercises, Iranian military leaders publicly revealed for the first time the existence of a secret network of what they characterized as 'underground missile silos'. These structures are distinctly different that those that have been built by the United States, Russia and China. The Shahab-3 is housed in a chamber that has a vertical tunnel that the missile flies through. Both the chamber and the tunnel are large relative to the missile diameter. This creates a very large volume that the rocket exhaust gases can expand into, greatly reducing the challenges associated with flying out of the types of ballistic missile silos built by the United States, Russia and China.

Iran's IRGC Aerospace Forces Commander BG Amir Ali Hajizadeh told state media that the Guard had the capability to "attack all American facilities in the region by these [Shahab 3] missiles."[84] In a televised segment another military official said that Iran had been building this network for the past 15 years. The *New York Times* reported Iranian State TV as saying these silos held medium and long-range missiles.[85] In June 2012, the UN said these missile silos, "which have been reported for a number of years, have not been confirmed to be operational."[86] Figure 5 shows pictures from Iranian television of what is reported as a ballistic missile in one of these underground facilities as well as a partially opened launch door.

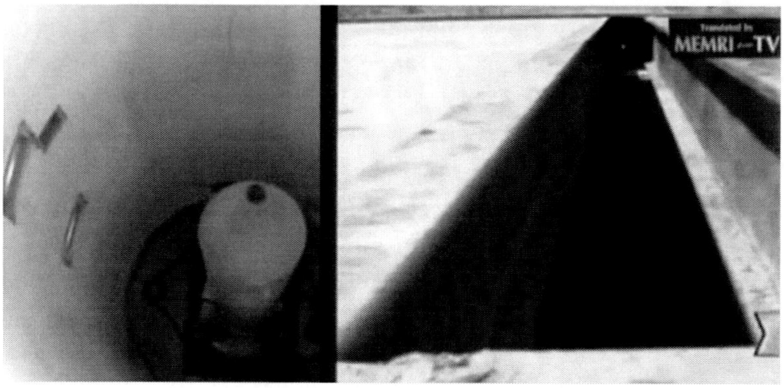

Figure 5. Iranian Launch Silo.

If the Iranian military official was quoted correctly, it would mean that work would have started on these silos around 1996. In fact, perhaps the first public account of Iranian interest in such came in 1993 when then Iranian Defense Minister Foruzandeh led an economic delegation to the DPRK and may have discussed the possibility of joint production of the Nodong missile and construction of underground missile shelters at 18 sites across Iran.[87] Construction may have been underway by 1998 when Israeli Prime Minister Netanyahu said Iran is "building an enormous infrastructure [including] hardened missiles silos," which can house Iranian ballistic missiles and protect them against U.S. or Israeli preemptive strikes.[88]

The next time there was any apparent mention of Iranian missile silos came in 2006. Former head of Israel's missile defense agency and missile engineer Uzi Rubin wrote that "there are indications they [Iranians] are now constructing fixed silo-like hardened sites to make their missiles even more survivable."[89] Rubin mentioned this in the context of Iran's MRBMs. It was not until 2008, however, that more specific claims about silo construction, hardened sites, locations, and imagery, began to surface in public.[90] Two different blogs, gemint.com and armscontrolwonk.com, wrote about the same time in early 2008 of a possible silo missile base near Tabriz, Iran. O'Connor (gemint.com) later wrote that:

> between 2003 and 2005, a silo complex was constructed on the grounds of an existing SSM [surface-to-surface missile] facility south of Tabriz in northwestern Iran. The complex consisted of two missile silos, likely intended to house Shahab-3 series IRBMs (sic) based on the existing facility. Between 2007 and 2008 a second pair of silos appeared roughly 700 meters SE of the initial pair. A third pair of silos appeared approximately 14 kilometers to the west near the town of Khosro Shahr.[91]

Source: CRS, Library of Congress.

Figure 6. Tabriz Ballistic Missile Silo Base 1.

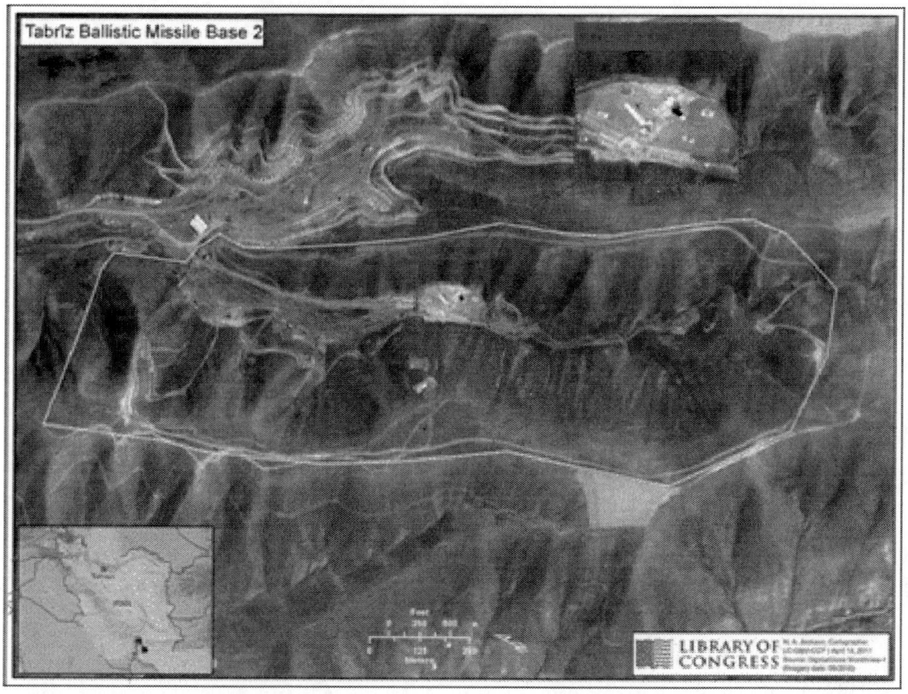

Source: CRS, Library of Congress.

Figure 7. Tabriz Ballistic Missile Silo Base 2.

Based on these previous assessments and using imagery obtained through the Library of Congress, this report locates what appear to be those missile silos in northwest Iran near Tabriz. See Figure 6, "Tabriz Ballistic Missile Silo Base 1," and Figure 7, "Tabriz Ballistic Missile Silo Base 2." Imagery obtained through the Library of Congress dated around 2000 and examining additional imagery dated about once every year or two at this location shows what appears to be gradual construction of the second silo site referenced by O'Connor. This second silo pair may have been completed by 2011 from looking at that imagery and is labeled 'Site 2' in Figure 6, "Tabriz Ballistic Missile Base 1."

Since then, only a few other experts have mentioned these silos until the Iranian videos and interviews in June 2011. To date, there does not appear to have been any official U.S. public assessment or acknowledgment of these silos. It is unclear what the absence of any such assessment might mean.

"Secret" MRBM Tests (2010-2011)

In comments to the British parliament in June 2011, British Foreign Minister William Hague said Iran had conducted three secret tests of ballistic missiles capable of carrying nuclear weapons in contravention of UN Security Council Resolution 1929.[92] An Iranian spokesman denied the British allegations, saying that none of the missiles tested by Iran has a nuclear capability.[93]

Britain apparently reported these tests to the UN, but had not previously made them public.[94] Although it received little media attention at the time, a report by a Panel of Experts at the UN became public the month before Hague's speech in London.

According to *AP* the UN report said Iran launched a liquid-fueled Shahab-3 missile with a range of about 800 kilometers, and one or two solid-fueled Sejil 1 missiles with a range of about 2,000 kilometers.[95] *AP* further said the UK believe these missile tests showed that Iran's leaders wanted to avoid scrutiny over "the real extent of their weapons programs."

This unpublished May 2011 report[96] was not circulated because it had not been approved by its parent UN Security Council, apparently for political reasons—China and Russia opposed its release according to UN sources. Nonetheless, the news it shed on previously undisclosed Iranian MRBM tests was highly significant.

Specifically, the report mentioned the launch of the Sejil/Ashura in October 2010 and a Sejil and Shahab-3 test in February 2011.[97] By early July 2011, Iran conceded the two February tests. *Reuters* reported that Iran state television admitted test-firing two long-range missiles from the "Semnan province into the mouth of the Indian Ocean" about 1,900 kilometers away sometime between January 21 and February 19, 2011.[98]

BG Hajizadeh, head of the IRGC's Aerospace Division, said U.S. spy planes were operating in the area where the missiles hit, but it was "interesting that they [the United States] did not report it."[99] *Jane's* said these medium-range missile tests were likely the Sejil-2 or the Shahab-3.[100]

The fact that there was neither public announcement by Iran at the time of these launches, nor any public condemnation from the United States or from any other any other nation represented a notable departure from the past. Three key questions are thus raised.

First, what might account for this change in Iranian policy that normally publicizes or televises missile tests? There are several conceivable explanations:

- One or more of the tests might have failed and Iran did not want that known. This could explain what happened with the first launch, but evidence is that the February flights were successful, or
- Iran may have tested to a new, untried range for which it did not want international attention. Only after media reports and the leaked UN draft report did Iran concede the longer range tests into the Indian Ocean; or
- Iran may have decided for whatever reason that it no longer wanted to publicly demonstrate its MRBM capabilities, perhaps over concern regarding UNSCR 1929. But in "Great Prophet" 7 Iran said it had flown a 2,000 kilometer missile, which should be noted was not independently verified. And Iran had declared two other Sejil launches and a Shahab-3 test in 2009, after UNSCR 1929 passed earlier that year. Also, just because Iran did not publicly announce or show these tests does not mean they cannot be verified by others.

A second key question is why neither the United States nor any other nation at the time chose to criticize Iran for those tests, which to many seemingly violates UNSCR 1929. Possible reasons might include:

- The United States might have wanted to forge a consensus behind closed doors for additional sanctions and worried that public attention to the tests might make it harder to bring in Russian or Chinese support, or both.
  U.S. Permanent Representative to the UN Susan Rice complained about Iran's noncompliance with UNSCR 1929 in June 2011 and referred to the draft Panel of Experts as containing "troubling findings, including significant evidence about several reported violations" of UN sanctions related to Iran, but implied some Member States were holding up public release.[101] Some have noted that the Final Report released in June 2012[102] did not mention the three tests from the unpublished earlier Final Report in May 2011. This was for no other reason than the missile launches occurred in a different mandated reporting period.
- Due to the sensitive, ongoing talks with Russia over the U.S. European Phased Adaptive Approach (EPAA), there may have been some reticence within parts of the U.S. Government to "shove this" into the Russian face. Keeping any announcement of these Iranian tests out of the public debate may have been seen as having a greater policy priority in order to facilitate an agreement with Moscow.

Intercontinental-Range Ballistic Missiles (ICBMs)

Traditionally, the United States has defined ICBMs as those ballistic missiles capable of achieving ranges greater than 5,500 kilometers (about 3,400 miles). To date, only five countries have deployed operational ICBMs (all with nuclear weapons): the United States, Russia, China, France, and the UK.[103]

Only the United States and the Soviet Union achieved those capabilities entirely indigenously; to some degree China, France and the UK received significant assistance. Some countries, such as Iran, are believed by some observers to have ICBM programs in varying stages of development.

Table 2. Iranian MRBMs (U.S. Government)

	2000	2001	2002	2003	2004	2005	2006	2007	2008	2009	2010	2011	2012
U.S. Intelligence Community Assessments*													
National Air and Space Intelligence Center (U.S. Air Force), Reports on Ballistic and Cruise Missile Threats**													
Shahab 3	Not yet deployed			< 20 launchers			< 20 launchers			< 50 launchers (all variants)			
Shahab 3 variant							Not yet deployed						
Shahab 4	Not yet deployed/ not flight tested												
New MRBM/ Shahab4?				Not yet deployed									
New MRBM							Not yet deployed			Not yet deployed			
Shahab 5	Not yet deployed/ n ot flight tested												
New MRBM/ Shahab5?				Not yet deployed									
IRBM/ ICBM							Not yet deployed			Undeter-mined			

Sources: *Annual Threat Assessments, 721 Reports. **NASIC (U.S. Air Force), Reports on Ballistic and Cruise Missile Threat. Missile designations as used by the source.

Notes: Where launcher numbers only are noted, actual missile inventory may be larger because launchers can be reused to fire additional missiles.

ICBMs, fitted with nuclear warheads, pose significantly greater challenges to U.S. national security than nuclear weapons alone. Director of Central Intelligence George Tenet testified in 2000, "these countries calculate that possession of ICBMs would enable them to complicate and increase the cost of US planning and intervention, enhance deterrence, build prestige, and improve their abilities to engage in coercive diplomacy."[104]

Since the late 1990s, the U.S. intelligence community has repeatedly stated, with a number of caveats, that Iran would be able to test an ICBM capable of reaching the United States by 2015, which is at least 10,000 km away. Other countries do not necessarily share this assessment and Iran has long denied that it is developing an ICBM capable of reaching the United States. Open-source analyses of ballistic missile tests in both Iran and North Korea raise questions about whether Iran may be developing a ballistic missile capable of reaching the United States or seeking to develop a space launch capability that could provide the technological basis for conversion to an ICBM. This section examines these issues.

Intelligence Estimates: U.S. and Others

The United States

In 1999, the U.S. intelligence community assessed that at some point the United States would probably face ICBM threats from Iran.[105] This remains the official U.S. position. More specifically, the 1999 assessment warned that "Iran *could test* an ICBM in the last half of the next decade using Russian technology and assistance" (emphasis in the original).

A similar report was issued in 2001.[106] Such a test did not occur in this time frame, but in 2010 the DIA assessed that, by 2015:

> with sufficient foreign assistance, Iran could develop and test an intercontinental ballistic missile (ICBM) capable of reaching the United States. In late 2008 and early 2009, Iran launched the Safir, a multi-stage space launch vehicle (SLY), demonstrating progress in some technologies relevant to ICBMs. Iran displayed its next generation SLY, the Simorgh, in February 2010. The Simorgh is much larger than the Safir and shows progress in booster design that could be applicable to an ICBM design.[107]

The 2012 annual report on Iran's military power to Congress (April 2012) restated that "Iran has launched multistage space launch vehicles that could serve as a test bed for developing long-range ballistic missile technologies."

These assessments are often interpreted as stating that Iran will have nuclear-tipped ICBMs capable of striking the United States by 2015, but the unclassified intelligence statements continue to place various caveats on potential capability to test at some date. These assessments focus only on the ability to test an ICBM and do not make any judgments about the ability of Iran to successfully deliver a nuclear warhead at ICBM range.

These intelligence statements serve as the official U.S. basis for assessing the Iranian ICBM threat to the United States and to its friends and allies. These assessments drive U.S. military efforts designed to respond to such threats, such as the U.S. BMD program in general and the U.S. missile defense system in Europe specifically, as well as U.S. diplomatic and other efforts such as sanctions to dissuade or slow down Iranian long-range ballistic missile programs.[108] However, they do not offer a probability assessment for such technological assistance being available.

These assessments do not mean that currently universal agreement exists within the U.S. intelligence community on the issue of an Iranian ICBM. According to these same unclassified statements, some within the intelligence community argued that an Iranian ICBM test was likely before 2010 (which did not happen), and very likely before 2015. Other U.S. officials believed, however, that there is "less than an even chance" for such a test before 2015. Furthermore, U.S. assessments are also conditional in that an Iranian ICBM capability would have to rely on access to foreign technology, from, for example, North Korea or Russia.[109] Finally, some argue that an Iranian ICBM could be developed out of the Iranian space program under which a space-launch vehicle might be converted into an ICBM program. In the 1990s, some argued that Iran could have developed and tested such a space launch vehicle by 2010. Iran successfully demonstrated a space launch capability in 2009 with the launch of a low-earth orbit satellite, but the IC has not assessed that Iran has conducted an ICBM test or acquired an ICBM capability.

Regarding the relation between ICBMs and space launch capability, a UN Security Council Panel of Experts "reached a consensus" in 2012 that

> both ballistic missile and space launch programmes shared a great deal of similar materials and technology, including systems for propulsion, control and navigation. The Panel also noted that, although some examples existed of ballistic missile programmes developed from space launch programs, in general there were more examples of the reverse – space launch programmes developed on the basis of ballistic missile programmes.[110]

Some observers argue that although the U.S. position may be based upon a realistic assessment, it is also a worst-case analysis of the potential threat from Iran. They argue that "with rare exception this level of threat has rarely turned out to be the historical reality."[111]

Other Governments and the UN

Some governments support the contention that Iran will be capable soon of developing ICBM capabilities, while others do not share that assessment. In 2008, a UK House of Commons, Foreign Affairs Committee report noted that Iran made no secret of its space launch aspirations, which "technology has some value for longer-range and intercontinental ballistic missiles. We believe Iran could test such systems by the end of the decade [presumably meaning 2010]. If it acquired a complete long-range ballistic missile system, it could achieve such a capability more quickly."[112] Although dated, it is worth mentioning here as indicative of a broader belief that Iran could develop an ICBM capability at some point.

In 2005, the conservative-leaning German tabloid newspaper *Bild* reported that, according to information obtained by the German Intelligence Service (BND), "Iran has bought from North Korea 18 assembly kits of the mobile BM-25 type [and that] Ahmadinejad has or had them upgraded on the basis of the Russian submarine missile SS-N-6."[113]

Russia, however, has long denied that any such transfer of R-27s or SSN-6s took place[114] and has indicated that an Iranian ICBM was not likely for the foreseeable future. In 2006, the Chief-ofStaff of Russia's Armed Forces said Iran did not have the technological potential to build ICBMs. Yury Baluyevsky said "according to our information, Iran possesses neither the technological nor technical resources to build a missile with an intercontinental range."[115] A few years later, U.S. and Russian technical experts and officials gathered to discuss Iranian programs. The Russians indicated they believed Iran

> was a long way from building intercontinental ballistic missiles that could hit the United States.... Iran lacks appropriate structural materials for long-range systems.... Iran can build prototypes, but in order to be a threat to the U.S. or Russia, Iran needs to produce missiles in mass quantities, and it lacks materials sufficient for the type of mass production needed to be a security threat. Russia further noted that the technology for longer-range missiles is sophisticated and difficult to master.[116]

Israel reportedly believes Iran will have ballistic missiles capable of striking the United States within two to three years. In recent remarks, Israeli Finance Minister Steinitz said the Israeli assessment was in line with a 2010 Pentagon assessment that Iran could build a U.S. missile with the range to strike the United States by 2015. He said, the Iranians "are working now and investing a lot of billions of dollars in order to develop ICBMs ... and we estimate that in two to three years they will have the first ICBMs that can reach the east coast of America. So their aim is to put a direct nuclear ballistic threat ... to Europe and to the United States."[117]

In 2012, a UN Security Council-approved Panel of Experts report that, among many other things, cast doubt on Iran's capacity to develop long-range ballistic missiles:

> While the Islamic Republic of Iran is actively producing its own missiles, it remains reliant on foreign suppliers for components, materials and equipment. According to some experts, there is no evidence that the Islamic Republic of Iran possesses the technology necessary to ... construct longer range missiles. It also appears that the Islamic Republic of Iran continues to import whole engines, or at least critical engine components, for its liquid-fueled missiles, and requires components for guidance systems.[118]

The View from Iran

Iran disputes that any of its ballistic missile programs under development would be capable of reaching the United States. Its most recent denial came when the Pentagon released its April 2010 report on Iran's Military Power. Among other things, that report reaffirmed previous intelligence assessments since the late 1990s that Iran could develop and test an ICBM capable of reaching the United States by 2015.[119] In response to this report, Iranian Defense Minister Ahmad Vahdi stated that "the Islamic Republic of Iran has no plans to build such a missile."[120]

Iran denies reports that it collaborates with other nations to develop ICBMs. In response to charges that Iran obtained ballistic missiles from North Korea based on the Russian R-27 SLBM design, Iranian Ambassador to Russia Mahmoud Sajjadi said in November 2010 that "Iran has received no [such] missile from North Korea ... [and] we do not need such military services from any country at the present time."[121]

The Commander of the Aerospace Division of the IRGC, BG Amir Ali Hajizadeh, said in 2011 that Iran "possesses the technology," but has "no intention of producing" missiles with ranges greater than 2,000 kilometers.[122] Hajizadeh further said Iran was interested only in the ability to target U.S. and Israeli bases in the region.[123]

Space Launch Program

Organization

The Iranian Space Agency (ISA) coordinates Iran's space activities. Iran is also one of the 24 founding members of the UN Committee on the Peaceful Uses of Outer Space, established in 1958, well before the 1979 Islamic Revolution. The ISA was established in 2004 and given responsibility to support all of Iran's activities concerning the peaceful uses of space. ISA falls under the leadership of a Supreme Council of Space that is chaired by Iran's President. This Council has policy making responsibilities regarding Iran's use of outer space, for building, launching and using satellites, approving all government and private sector space programs, and identifying areas of potential regional and international cooperation. ISA is affiliated with Iran's Ministry of Communication and Information.

Iran conducts its space launch activities primarily out of the Semnan Launch Center in the north central part of the country. Iran has reportedly launched previously from an area near Qom and from Emamsharh as well. From these locations Iran is able to send its rockets in a southeastern direction over the entrance of the Persian Gulf and then over the Indian Ocean. Figure 8, "Space Launch Vectors (Current)," shows these launch corridors, which are somewhat constrained if Iran wants to avoid overflying other countries such as Oman and Pakistan. The slightly darker lines extending out a couple hundred kilometers show where the 1st stage rocket motors might be expected to fall to earth.

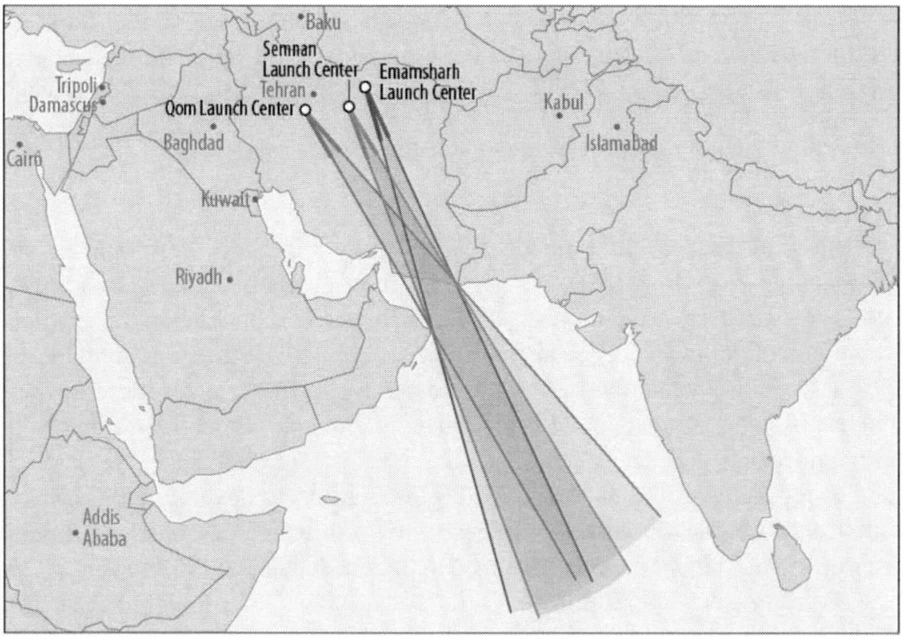

Source: CRS.

Figure 8. Space Launch Vectors (Current).

Iran announced in December 2010 that it would build a new space launch complex in the Sistan and Baluchistan Province, which is at the entrance of the Persian Gulf, and which Iran's Defense Minister Vahdi indicated in June 2012 was about 80% complete although Vahdi gave no clue as to the center's location.

Some experts believe a new space complex, first mentioned in December 2010, is being built in southeastern Iran in the Sistan and Baluchistan Province.[124]

The fact that such a complex is so close to the Gulf and highly vulnerable to attack suggests its purpose is for space launch activities. Figure 9, "Space Launch Vectors (Prospective)," illustrates how significantly enhanced and more flexible Iran's space launch endeavors would be increased over what they are currently with this new space complex. Range safety concerns would similarly be greatly alleviated by having rocket stages and possible failed launches occur over wide swaths of the ocean rather than land. From this southern launch location, a more southeastern trajectory also provides some advantage of the earth's rotation in achieving orbit.[125]

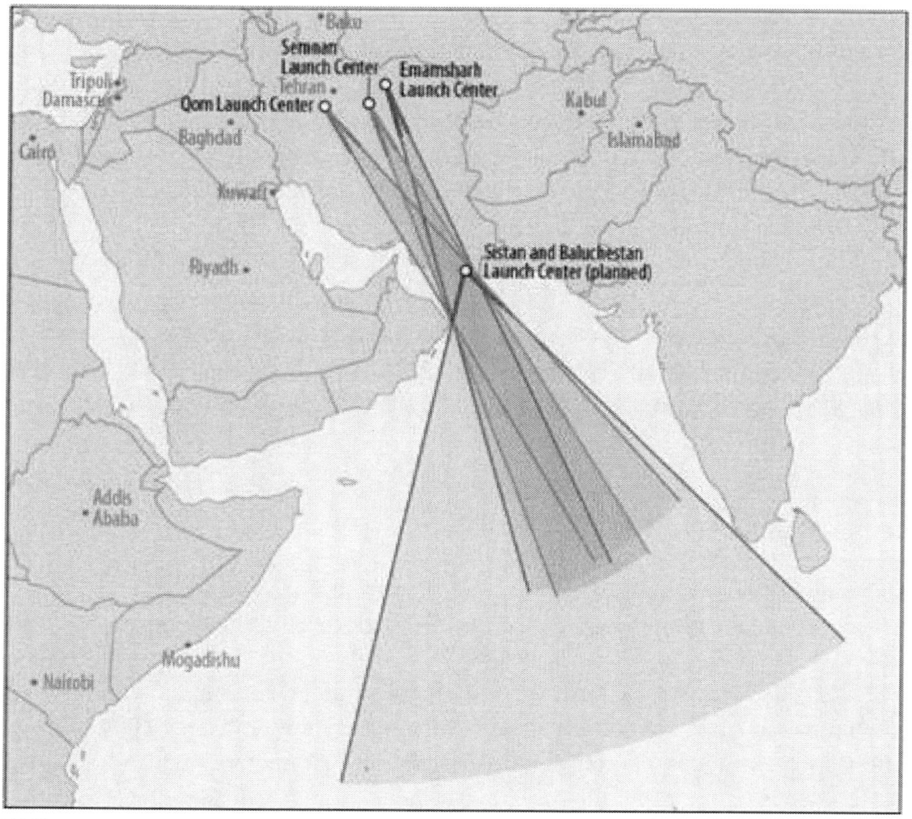

Source: CRS.

Figure 9. Space Launch Vectors (Prospective).

Space Launch Vehicles

Iran used the Safir SLV to launch the Omid satellite into orbit in February 2009. The Safir SLV used a first stage based on the single stage Shahab 3 with a specially designed second stage. Iran continued to develop this liquid-fueled SLV. The Safir is designed to carry a light payload into low-earth orbit and as such is not considered to be capable of providing a long-range nuclear weapons capability. The Safir-2B is a second generation Safir SLV designed to put a 50 kilogram payload into low-earth orbit.

A more powerful rocket was displayed as a mockup in 2010 and named Simorgh. It appeared to be designed to carry heavier Iranian satellites into orbit using four main engines. Iran claims the Simorgh could put a 60 kilogram payload into low-earth orbit. Comments made by Iranian officials suggest the design has since undergone various modifications. Simorgh (which Iran now calls the Safir-2) was to have first flown in 2010, but is now expected to be launched at either a newer site at Semnan or the newly planned space complex in southeastern Iran.

Orbital Launches

Satellite Launch Effort to Date

Iran became the ninth country[126] to demonstrate an indigenous space launch capability on February 2, 2009 when it launched an Omid satellite from a Safir 2 rocket. The launch occurred at the Semnan Missile and Space Center east of Tehran and flew southeast over the Indian Ocean.[127] Space-track.org confirmed soon thereafter that two objects had entered low-earth orbit. The Omid satellite reentered the atmosphere on April 25, 2009 and fell into the south Atlantic. The Safir 2 rocket body reentered the atmosphere on May 31, 2009. Iran said that Omid was a communications satellite.[128]

U.S. officials, however, objected to the Iranian space launch. The Department of State said: "Iran's ongoing efforts to develop its missile capabilities remain a matter of deep concern. Recently, Iran's development of a space launch vehicle (SLV) capable of putting a satellite into orbit establishes the technical basis from which Iran could develop long-range ballistic missile systems."[129] A spokesman for the Department of Defense said the launch was

> certainly a reason for us to be concerned about Iran and its continued attempts to develop a ballistic missile program of increasingly long range. Although this would appear just to be the launch of a satellite, their first, obviously there are dual-use capabilities in the technology here which could be applied toward the development of a long-range ballistic missile. And this is a cause of concern to us, and I think to certainly everybody in the region – Israel and their Arab neighbors – as well as to our allies in the region.[130]

Various analysts have argued that to some degree Iranian missile experts have modified or reverse-engineered older Soviet-era rockets in Iran's space program. They point out how identical they look, how comparable their performance parameters are, and so on. Despite compelling evidence supporting their claims, others strongly disagree. German scientists Schmucker and Schiller maintain that "reverse engineering is so difficult that there is not one single proven example for successfully reverse engineered missiles and rockets."[131]

Iran has launched two additional satellites. The Rasad-1, reportedly an imaging satellite, was successfully launched on June 15, 2011. The Navid-e Elm-o Sanat was reportedly successfully launched on February 3, 2012; it remained in orbit for about two months. The Navid is an experimental student-built satellite designed to test camera and telecommunications equipment.

Planned Satellite Launches

In January 2011, Iranian Defense Minister Vahdi announced that two satellites were being readied, including the Fajr satellite and the Rasad satellite mentioned above, which were capable of capturing imagery with a resolution of more than 200 meters. He also said a space laboratory called Explorer-4 was being readied.[132]

On February 7, 2011, President Ahmadinejad named four satellites to be launched: Fajr, Rasad, Amir-Kabir-1 and Zafar.[133] Table 3, "Iranian Satellites (Current and Planned)," provides an overview of current and planned Iranian satellite launches.

Sub-orbital Launches

As part of a nation's space program, it is common to develop, test and launch a number of rockets into space for short durations to measure the effects of space and study its effects on animals in preparation for subsequent manned space flight. Iran has outlined such a plan and has moved forward with those plans.

This is another indication that Iran intends to pursue advanced space flight in the future.

Table 3. Iranian Satellites (Current and Planned)

Name	Purpose	Date
Sinah-1 (Iranian design, Russian built)	Reconnaissance	Launched by Russia on Oct. 28, 2005
Environment-1 (Iran, China, Thailand)	Joint research; earth observation	Launched by China on Sept. 6, 2008
Omid	Research and Telecommunications	Launched by Iran on Feb. 2, 2009
Fajr	Imaging with experimental GPS system	Delayed, planned for fall 2012
Navid-Elmo-Sannat or Ya Mahdi (experimental student satellite)	Telecommunications and reconnaissance	Launched by Iran, February 3, 2012
Rasad	Imaging	Launched by Iran, June 15, 2011
Mesbah (Iranian-Italian collaboration)	Telecommunications	Satellite being held by Italy because of sanctions
Mesbah-2	Limited telecommunications	Planned for 2011-2012
AUT-SAT (experimental student satellite)	Remote Sensing	Planned for 2011-2012
Zafar-1	Imaging	Planned for 2012
Zohreh	Communications	Planned for 2014
Qaem	Communications	Planned for 2016
Toloo	Reconnaissance with SIGINT capabilities	Launch date not confirmed
Pars Sepehr	Remote sensing	Launch date not confirmed
Pars-2	Remote sensing	Launch date not announced
Iran-APSCO (Asia-Pacific Space Cooperation Organization)	Design, build and launch 10 research, remote-sensing and telecommunications satellites	Launch dates not confirmed
Besharat (Iran-Organization of the Islamic Conference)	Not announced	Not announced

Sources: Various Iranian and international news sources and *Jane's Intelligence Review*, April 4, 2011.

On February 25, 2007 Iran reportedly launched a rocket, but no information on what equipment or cargo it carried was provided. A year later on February 4, 2008 Iran said it launched a suborbital rocket named Kavoshgar-1, which reportedly transmitted various scientific data on the earth's atmosphere. Kavoshgar-2 launched in November 2008 carrying a small space laboratory. A third rocket, Kavoshgar-3 launched on February 3, 2010 with several animals (a rodent, several worms, and two turtles) and transmitted live video of the small environmental lab. Iran thus became the sixth nation or consortium to send animals into space.[134] Kavoshgar-4 launched on March 15, 2011. Although designed to carry a monkey, this particular launch did not include any animals. In September 2011, Kavoshgar-5 was launched with a monkey on board in a test capsule, but the rocket failed. Later in 2011, Iran suspended its sub-orbital flight program for further review, but in May 2012 Iran announced the program to send live animals into space would resume shortly. The Kavoshgar-1 was apparently a modified Ghadr-1; the other Kavoshgars likely were based on either Zelzal or Nazeat rocket motors.

Manned Space Flight

Iran first expressed interest in manned space flight during a Soviet-Iranian Summit on June 21, 1990. Both Presidents agreed in principle to conduct joint Soviet-Iranian manned flights to the Soviet Mir space station. This goal was not realized because of the break-up of the Soviet Union soon thereafter.

It was not until 2008 that Iran announced plans for its own manned space program, as well as plans for the development of its own spacecraft and a space laboratory. [135] The head of Iran's Aerospace Industries Organization added that Iran intended to launch a manned mission into space within a decade. He described this as Iran's top priority for the next decade in order to make Iran the leading space power of the region by 2021.[136]

In August 2010, Iranian President Ahmadinejad said Iran's first astronaut would be sent into space on an Iranian space craft no later than 2019. This would be preceded by an Iranian manned sub-orbital spaceflight by 2016 in preparation for the eventual orbital spaceflight. There has also been some unconfirmed news regarding Iranian participation in a Chinese space station.

The fact that Iran has signified a commitment to a public, manned space effort over the next decade is indicative of the kind of space launch program one would expect from a country pursuing a legitimate program. Although this does not mean Iran will not learn much from a space launch program that it could apply to long-range ballistic missiles, some might argue that its use of personnel and resources in this way significantly weakens the case that Iran's space program is merely rhetorical masking of a covert ICBM program.

FOREIGN-IRANIAN COOPERATION

Since the early 1980s, many thousands of articles and news stories have been written about Iran's efforts to procure ballistic missiles, components and materials, and to develop an indigenous missile and space launch infrastructure. A considerable number of these items provide details of interest. Although many of those articles are likely accurate, it is not entirely possible to sort them all out from those accounts that are not entirely accurate. On the other hand, official public U.S. and other government sources generally do not provide such specificity, but instead are very useful to track broader trends over time.

This section is included to provide an overview from such public official assessments because no such overview is found elsewhere probing the extent of foreign cooperation with Iran's ballistic missile and space launch programs. Furthermore, this section describes not only the degree to which Iran developed close working relationships with others on the way to developing its own indigenous capabilities, but also demonstrates that these relationships largely continue out of necessity for Iran.

Around 1982, Iran began to focus on developing an indigenous solid-propellant rocket industry, but turned subsequently to other countries in varying degrees to acquire whole systems, components and training to build up its own ballistic missile force and develop its own production capabilities. Iran's efforts in this regard evolved and perhaps even diminished over time with respect to specific countries. As a result of this decades long effort, Iran is now considered to have become self-sufficient in major parts of its own ballistic missile programs, yet it is still apparently reliant on outside sources for some key missile components, especially for further development of its longer range ballistic missile program.

The Libyan and Syrian Relationship

Perhaps the earliest evidence of direct foreign missile cooperation with The Islamic Republic of Iran began in the early 1980s.[137] Iran reportedly began secret negotiations with Libya in 1984 to acquire short-range Scud-B ballistic missiles and in the following year received 20 Scud-Bs and 2 TELs (transporter erector launcher).[138] In 1986, Iran apparently received about a dozen Scud-Bs from Syria. It does not appear that Iran acquired additional Scuds from Libya or Syria after those transfers.

North Korean-Iranian Cooperation

Iran's relationship with the DPRK in developing ballistic missiles has been of long-standing. Starting with acquisition of Scud missiles from the DPRK in the 1980s, Iran has developed a close working relationship with Pyongyang in many if not all of its ballistic missile programs that subsequently led to Iran becoming increasingly self-sufficient in those programs.

Currently, Iran has likely exceeded the DPRK in its technical capacity to develop, test and build ballistic missiles.[139] But Iran may, to some extent, remain reliant on others, perhaps including the DPRK itself, for certain materials in the production of its ballistic missiles.

Then-Director of Central Intelligence Robert Gates testified in 1992 that North Korea had sold other countries longer range missiles and the technology to produce them.[140] The following year, then-Director of Central Intelligence R. James Woolsey[141] added that North Korea has sold Syria and Iran extended range Scud Cs and apparently agreed to sell missile technology prohibited by the MTCR. The IC further expressed concern that the DPRK was developing and marketing a new 1,000 km range missile and apparently had no threshold governing its sales; it appeared willing to sell to any country with the cash to pay. Already, North Korea was selling two versions of ballistic missiles—the normal modified Scud (500 km range), which they sold to several countries in the Middle East. They also had a number of discussions and contract talks for the 1,000 km missile, Woolsey testified. In follow-up questions for the record, the IC later wrote that Iran, "one of North Korea's best customers for ballistic missiles and related technology, is likely to be one of the first recipients of the 1,000 km No Dong. By the end of this decade [1990s], Iran could be able to assemble short-range (Scud B and Scud C) and medium-range No Dong ballistic missiles."[142]

Woolsey also commented on a report by the Russian intelligence community: "the IC has reviewed the Russian report on nonproliferation and finds it to be a credible unclassified document that generally corresponds to Western appraisals."[143] That Russian report said Iran's missile potential was confined to Scud B SRBMs received from Syria and North Korea. Using technology obtained from Egypt, North Korea was upgrading the Scud-class missiles purchased earlier from the USSR and exporting them to countries such as Iran. The same Russian report noted that North Korea's "Nodong 1 IRBM," which the DPRK intends to offer on the world market, was at the testing stage.[144]

The 1990s saw several recurring trends 1990s between Iran and North Korea as assessed by the IC in its annual threat assessments before Congress. First, North Korea's ongoing export of ballistic missiles provided a qualitative increase in capabilities to countries such as Iran. Second, Iran was using its relationship with North Korea, and the significant amount of ballistic missile goods and services it provided to Iran, as an important means by which Iran was achieving its goal of self-sufficiency in the production of MRBMs. Third, Iran could significantly shorten the acquisition time for an ICBM by purchasing key components or entire systems from potential sellers such as North Korea, including potentially significant inputs of space launch vehicle technology and support.

It was not until 2006 that Iran publicly acknowledged for the first time that it had obtained missiles from North Korea during its war with Iraq, but added that it no longer needed Pyongyang's assistance: "We received these [Scuds] from foreign countries like North Korea but 17 years after the war we were able to design all of these pieces and even their fuel," said the chief commander of the IRGC.[145] Some observers cast doubt on Iran's ability to produce even the Scud Bs and Scud Cs (Shahab-1 and Shahab-2 respectively) entirely indigenously without any foreign support for materials or components, or both, even through today.

In the latter 2000s, the IC continued to provide testimony on Iran's longstanding relationship with North Korea with respect to the purchase and development of ballistic missile technology. The IC continued to assess that North Korean cooperation was ongoing and significant to Iran's ballistic missile programs.

The Russian-Iranian Relationship

According to official public accounts, since the early 1990s, Russia's involvement with Iranian ballistic missile programs evolved slowly at first. But whether the Russian government itself was officially or indirectly involved with non-government Russian entities and Iran on these matters remains unclear. Whatever the case may be, the involvement of Russian individuals and entities expanded despite Russian government efforts to curb such involvement, leading to the acceleration of some Iranian missile programs. In the last few years, Russian involvement appears to have receded. From 2004-2007, Russian arms transfer agreements with Iran totaled about $1.6 billion (current U.S. dollars); from 2008-2011 total arms transfer agreements totaled about $100 million (current U.S. dollars).[146]

During the era of the Soviet Union, it does not appear that Iran acquired any ballistic missile systems, technology or training from that country. As the Soviet Union broke up, however, U.S. officials began warning that the former Soviet Union could become a source of proliferation concern. Then-Director of Central Intelligence Robert Gates testified in 1992 that specialized Soviet-era defense industries facing cuts in military funding might turn to the international marketplace where pursuit of hard currency could take precedence over proliferation concerns.[147]

Gates repeated these concerns in 1993,[148] and his successor, R. James Woolsey testified that Russia and Ukraine were showing a growing willingness to sell missile technology prohibited by the MTCR (Missile Technology Control Regime) because of internal nationalist and economic pressures.[149] Woolsey added that most reports of any such transfers to that date were hoaxes, exaggerations or scams, but that a few transfers of sensitive technology had occurred to China. He warned further that China in turn could pass more advanced Russian or Ukrainian military technologies to others. Although the potential for Iranian-Russian ballistic missile cooperation existed, the U.S. IC made no mention of any such relationship in 1994[150] or 1995.[151]

The Iranian-Russian relationship had changed by 1996, however. Congress was notified in the annual 721 Report[152] that Russia had become a "primary source for missile related goods" and had "supplied a variety of ballistic missile goods during the reporting period to foreign countries, especially Iran."[153] The 1997 annual threat assessment from the Intelligence Community stated that Iran acquired Kilo-class submarines from Russia and was upgrading its anti-ship missile capabilities.[154] Additionally, the 721 Report stated "that entities in Russia and elsewhere continued to supply missile-related goods and technology to Iran. Iran is using these goods and technologies to achieve its goal of becoming self-sufficient in the production of MRBMs. The foreign equipment obtained in 1997 will be essential for achieving that goal."[155]

Russian officials, apparently concerned about these activities, issued a broad decree in January 1998 prohibiting Russian companies from exporting items known or believed to be used for developing WMD or their delivery systems, whether or not those items were on Russia's export control list.[156] Acknowledging that Russian efforts had shown some signs of slowing proliferation activities, the U.S. IC noted the difficulty in restricting technologies with dual civil and military applications, especially when countries such as Iran could take advantage of shortcomings in Russian export controls.[157] More specifically, the 1998 721 Report said that Iran's success in -gaining technology and materials from Russian companies accelerated Iranian development of the Shahab-3 MRBM, which was flight tested in July 1998. The Report added that following intense and continuing engagement with the United States, "Russian officials have taken some positive steps in curtailing proliferation activities."[158]

Despite Russia announcing additional, new export controls, the IC reported in 1999 that "expertise and material from Russia continued to assist the Iranian missile effort in areas ranging from training, to testing, to components. There is no doubt this will play a crucial role in Iran's ability to develop more sophisticated and longer range missiles."[159] The first of two 721 Reports in 1999 stated that entities in Russia and elsewhere continued to supply considerable ballistic missile-related goods and technology to Iran, whose goal was to become self-sufficient. Despite steps taken to curtail proliferation activities, the Russian Government's "commitment, willingness, and ability to curb proliferation-related transfers remains uncertain. Moreover, economic conditions in Russia continued to deteriorate, putting more pressure on Russian entities to circumvent export controls."[160] The second of the two 721 Reports for 1999 reiterated much of the previous report and added that Russian entities provided substantial missile-related technology, training, and expertise to Iran that "almost certainly will continue to accelerate Iranian efforts to develop new ballistic missile systems, especially longer range ballistic missiles."[161] These assessments aligned with the 1999 National Intelligence Estimate (NIE) that stated Iran could test an ICBM that could deliver a several-hundred kilogram payload to many parts of the United States in the last half of the next decade (i.e., the late 2000s) using Russian technology and assistance.[162]

Russian entities continued to engage in transfers of missile related technology, goods and technical know-how throughout the 1990s until at least 2010. During this period, Iran moved increasingly toward self-sufficiency in the production of ballistic missiles. However, Iran certainly remains dependent on foreign suppliers for some key missile components," such as Russian suppliers (among others) according to the IC.[163]

A limited but significant level of ongoing cooperation is seen in each of the 721 Reports provided to Congress during the 1990s until the last several years. Also, a number of the IC's Annual Threat Assessments provided information about ongoing proliferation activities and concerns over the cooperation of some Russian entities with Iran. Through this period too, various Russian entities and individuals were sanctioned by the United States for their proliferation activities with Iran.

A change may have occurred in the last couple of years. Neither the 2011[164] nor the 2012[165] annual threat assessments to Congress indicated ongoing cooperation between Russian entities with Iranian ballistic missile programs. And the 2011[166] and 2012[167] 721 Reports say the same thing that "Iran continued to move toward self-sufficiency in the production of ballistic missiles, but almost certainly remains dependent on foreign suppliers for some key missile components. Entities in China and Russia along with North Korea are among likely suppliers." In other words, the specificity of earlier Russian cooperation is now more ambiguously described as collaboration with other countries and other foreign entities.

In the open-source literature, there are a significant number of reports, articles, news accounts and assessments that touch on Russia's involvement with Iran's ballistic missile programs. Although this body of literature can be speculative and misinformed, some of it appears well grounded, albeit based on unclassified technical or other assessments. Because of this, the public policy debate can be limited in terms of what can be considered informed and accurate, and what is inaccurate or misleading.

From some of the expert community, at least two slightly different schools of thought exist as to what Russian involvement in Iran's ballistic missile programs may have consisted of. The official, unclassified record discussed above does not necessarily rule out or definitively show that either view is represented accurately by the open-source literature.

One view suggests that Iran may actually have succeeded in gaining access to a licensed production line from Russian defense firms via North Korea. Noted German rocket scientists Robert Schmucker (who served as a weapons inspector in Iraq) and Markus Schiller have argued that North Korea did not indigenously develop its own ballistic missile capabilities, but rather procured Russian ballistic missiles that they then may have modified with technical assistance from entities elsewhere, principally Russia and perhaps China, with likely funding support from Iran. This licensed production arrangement, they argue, may continue today. Some of those systems in turn may then have made their way into Iran; these authors, like many others, believe many of Iran's ballistic missiles are of North Korean origin or manufacture.[168] Schmucker and Schiller assess that the Shahab-1 SRBM is identical to the DPRK Scud B, the Shahab-2 is almost identical to the North Korean Scud C, and the Shahab-3/No Dong is strongly related to the Scud B. Further Iranian adaptations of these missiles are evident, they argue. A second school of thought largely posits that Iranian scientists and technicians have been able to reverse engineer variations of derivatives of Soviet-era ballistic missiles with varying degrees of Russian assistance. How Iran actually acquired such technology is further debated, but DPRKIranian collaboration on reverse engineering is often suggested, with varying levels of assistance from Chinese entities as well. Michael Elleman[169] for instance, argues there is no evidence to suggest that Iran itself succeeded in procuring a licensed production line for liquid-fueled engines. Instead, Elleman has said it is possible but uncertain that Iran received Russian assistance in attempting to reverse-engineer and manufacture the North Korean No Dong engine, which is widely believed to be an adaptation of the Soviet-era Scud missile. MIT Professor Ted Postol wrote a detailed and significant contribution to the technical debate over Iranian ballistic missile programs.[170] He largely agrees that the Shahab series are of North Korean derivation and with some degree of reverse engineering they were adapted in Iran. Although Postol agrees there was likely Russian assistance, he argues that Iranian scientists and engineers have demonstrated their own significant expertise in their ballistic missile programs. Again, neither view may be entirely correct or entirely wrong. Both may have some correspondence with classified technical assessments as to how precisely Iran acquired and developed its ballistic missile capabilities with respect to Russian involvement.

Chinese-Iranian Cooperation

China has been involved in proliferating WMD and ballistic missile programs to other countries for a long time, and has thus proven a major concern for the United States. Although many U.S. policymakers deem China's progress perhaps too slow to be acceptable, others would argue that progress toward U.S. nonproliferation objectives is being made. Where once China exported entire systems, production capabilities, and key components often with direct or tacit government approval, many believe today that China no longer does those things. Instead, China has adopted its own export controls modeled after those of the MTCR and works with the United States to help stop some proliferation activities within China. This has now forced countries such as Iran to find and exploit weaknesses in the Chinese system, looking for lax enforcement and shopping around for PRC entities willing to export perhaps out of view of the Chinese government. From 2004-2007, Chinese arms transfer agreements with Iran totaled about $300 million (current U.S. dollars); from 2008-2011 total arms transfer agreements dropped to less than $50 million (current U.S. dollars).[171]

The U.S. Intelligence Community expressed concern in the early 1990s over China selling other unnamed countries longer range missiles and the technology to produce them, especially those technologies that might grow out of China's preexisting military cooperation with countries such as Russia and Ukraine. The IC pointed out that although China had agreed to observe MTCR guidelines when the United States lifted sanctions, unless sales of such missiles were actually stopped, it was likely these delivery systems would be mated with WMD capabilities, especially in the Middle East.[172] During this period it was also clearly a source of concern that China might be cooperating on missile programs with North Korea who might in turn assist Iran in its ballistic missile programs. Russian intelligence to this point in 1993, however, assessed that Iran's missile programs were restricted to Syrian and DPRK Scud purchases. According to Russia, available data confirmed there were "bottlenecks" throughout the whole of Iran's missile program, primarily a shortage of skilled personnel, science-intensive technology, scarce starting materials, and, possibly, the requisite amounts of financing.[173] In 1996, the Intelligence Community began to testify openly that China was among several countries and private consortiums that were willing to sell missile systems and technologies to developing countries around the world.[174] Although Iran was not named specifically in the annual threat assessment, the IC reported to Congress a more specific role between China and Iran in the 721 Report that year, stating "China and Russia have been primary sources for missile-related goods" to Iran."[175] Iran's efforts to acquire foreign missile technology was characterized as "unrelenting," and the Chinese were described as having provided a tremendous variety of assistance to both Iran's and Pakistan's ballistic missile programs. This charge was stronger the following year when the IC reported that Iran was using these goods and technologies (from several countries including China) "to achieve its goal of becoming self-sufficient in the production of MRBMs. The foreign equipment obtained in 1997 will be essential for achieving that goal."[176] It was not clear how key China's participation was compared to the other countries named: Russian and the DPRK.

By 1998, the IC testified that PRC defense industries were under increasing pressure to become profit making organizations – an imperative that put them at odds with U.S. interests.[177] The IC added that conventional arms sales had lagged in recent years, encouraging Chinese defense industries to look to WMD technology, primarily to Pakistan and Iran, in order to recoup any losses. Further, there was no question that China contributed to WMD advances in these countries. Moreover, although China had passed comprehensive laws governing nuclear technology exports, China's relations with some proliferant countries were long-standing and deep.[178] The 721 Report added little more, other than that Chinese entities provided a variety of missile related items and assistance to several countries of proliferation concern, implying, but not naming, Iran.[179] Since 1999 through today, in its annual worldwide threat assessments and through the 721 reports to Congress, the IC regularly discussed concern over ongoing Chinese or Chinese entity involvement with Iranian and other country missile programs,[180] despite some level of PRC government commitment to restricting such exports. From these assessments, China's role has apparently not changed significantly one way or the other. In part, China has disagreed with the United States and others with interpretation of some of its MTCR commitments to restrict missile-related exports. The PRC might therefore argue that the types of exports its companies are selling to countries such as Iran are legitimate. The U.S. position has been otherwise. But there might also be another explanation: lax PRC enforcement of its commitments and its laws. In 2011, the IC reported that

> Chinese entities – including private and state-owned firms – continue to engage in WMDrelated proliferation activities. The United States in recent years has imposed sanctions on several Chinese companies for sales of WMD- and ballistic missile-related technologies to states of concern. Although China has export control legislation that approximates MTCR controls, enforcement continues to fall short. Chinese entities

continue to supply a variety of missile-related items to multiple customers, including Iran, Syria, and Pakistan.[181]

Missile proliferation or problematic missile-related exports from China was not mentioned in the most recent IC annual threat assessment (2012).[182]

Net Effects on Iranian Programs

Iran has become self-sufficient in many aspects of its ballistic missile and space launch programs. But as noted in the above section, the U.S. IC has said on several occasions that Iran almost certainly remains dependent on foreign suppliers for some key missile components. The United States has not identified what these components are. According to an unreleased report by a Panel of Experts at the UN,[183] however, a number of specific items that Iran continues to procure from foreign sources indicate a lack of indigenous capability. Iran continues to seek navigation guidance units, including gyroscopes and accelerometers, testing and satellite navigation equipment, control systems, tracking telemetry equipment, transmitters, receivers, on-board sensors and transducers. In addition to these missile components, Iran lacks access to some high quality production materials to improve the effectiveness and precision of their missiles. These dual-use materials include carbon-carbon materials,[184] structural materials,[185] polymeric substances,[186] ammonium percholorate[187] and aluminum powder.[188]

Perhaps the most consequential contribution has been from China, mostly in the provision of solid-propellant production facilities and equipment and likely training. What China has provided and the accumulated knowledge gained by Iran in operating these facilities has likely enabled Tehran to build whatever size and range solid fuel missile they want, although this will require time and money.

Today Iran is viewed increasingly as a rising proliferator. Perhaps the irony of a restrictive nonproliferation regime is that Iran became largely self-sufficient. This likely took longer than Iran had wanted, but with a considerable indigenous capability Iran may now be positioned to export some of its own ballistic missile systems, components or knowledge to others. It has been noted that Syria continued work on establishing a solid-propellant rocket motor development and production capability with help from outside countries such as Iran.[189] But civil war in Syria raises doubts as to whether this will result in anything of consequence.

Challenges Today

Iran has demonstrated its willingness over decades to pursue the development, testing, acquisition and deployment of ballistic missiles as an essential part of its regional war fighting, deterrent and retaliatory capability. A determined adversary such as Iran has not shown that it is deterred or dissuaded by U.S. conventional military superiority, or by U.S. and international sanctions, or by the deployment of U.S. BMD capabilities.

Iran has similarly demonstrated resolve and commitment to a genuine space launch program as a matter of national pride and self-sufficiency in space in the face of widespread international condemnation. Just as every other space faring nation, however, Iran too will use space for a range of military purposes, such as for reconnaissance and communications.

Is slowing down Iran's ballistic missile and space launch programs, making them more difficult and expensive, and forcing Iran to find less reliable and capable alternatives an adequate solution for dealing with Iran? On this point there are significant disagreements and no consensus on how to proceed.

At the same time, is it realistic to believe that ending Iranian interest in acquiring ballistic missiles or pursuing a space launch capability is even possible? Some point out that other countries have given up their missile programs entirely, or forsworn ballistic missiles while developing an effective space launch program alone. But would Iran change its pattern of behavior short of a change in regime? Most are skeptical.

So this seemingly leaves the United States and its allies, many might reluctantly agree, with few better choices than the current apparent path. On the other hand, there are perhaps early indicators that Congress' attitudes regarding the reliability of the "assured destruction" deterrence policy of the Cold War era may be changing under the circumstance of both nuclear and missile delivery system proliferation. Among the pieces of evidence suggesting this change in attitude is Declaration 2 in the Resolution of Ratification accompanying New START.[190] Such a change in attitude might imply that defensive options will no longer be seen as 'destabilizing', as was the dominant view during the Cold War. Although the behavior of rogue states such as Iran may be catalyzing this change in attitude, the language of Declaration 2 also carries the strong implication that in the future Congress will gauge the missile threat more broadly than just focusing on the activities and progress of ballistic missile and nuclear programs in such rogue states. This is because Declaration 2 challenges even a mature missile power such as Russia to adopt a more defensive strategic posture in order to provide a political and policy basis for devaluing offensive missile and nuclear weapon programs in a broader grouping of threatening and potentially threatening states. In this case, the focus on assessing the progress of ballistic missile and nuclear weapons programs of rogue states might become less intense in Congress because responsive U.S. policies will become more broadly based. In short, specific actions by Iran may become a less powerful driver of U.S. strategic deterrence policy than is the case today. For the time being, however, Congress will likely continue to take a focused interest in Iran's role in the region and in its missile programs. Beyond efforts to impose increasingly stringent sanctions, it is unclear whether there is more Congress itself can do to affect Iran's commitment to its ballistic missile and space launch programs. But continuing to work with the Administration through cooperation and oversight, to stay prepared to take additional action as needed is a process many decision makers appear prepared to be committed to.

APPENDIX A. THE "GREAT PROPHET" EXERCISES (2006-2012)

Since 2006, Iran has conducted various military exercises that included some number of ballistic missile tests. These are called "Great Prophet" exercises by Iran and sometimes "Noble Prophet" or "Holy Prophet" by others.

Iran tests missiles at other times as well, but these particular missile tests are held in conjunction with other conventional land, sea, and air military exercises. From open sources it is neither possible to identify with precision the number and type of all missiles launched during these exercises, nor to identify all other test launches by Iran at other times.

Nevertheless, these ballistic missile exercises provide considerable information to U.S. and other intelligence communities that assess Iran's ballistic missile programs. Additionally, these exercises serve several purposes for Iran, including testing and evaluating weapon system performance, providing operational training and evaluation of military personnel, broadcasting Iran's capabilities to the United States and others throughout the Middle East, and showcasing military prowess to Iran's own populace. Many of these missile launches are shown in photos and video on Iranian television and other media and can be found on such places as YouTube.

There does not appear to be any single source that tracks or details the missile launch components of all these exercises. Therefore this appendix is added to this report.

Great Prophet 1 (March-April 2006)

Iran conducted military exercises in the Straits of Hormuz from March 31 to April 6, 2006. Most reports at the time were vague or conflicting about precisely what missile launches were involved. Various Iranian media reported launches of a medium-range radar-evading anti-ship missile, a different multi-warhead missile and a very fast torpedo.

Most media reports at the time, however, reported these simply as missile launches. Various observers outside Iran greeted Iranian technical claims with some skepticism or believed that these were not Iranian weapons, but more likely Russian and Chinese in origin. A Pentagon spokesman responded to the launches saying while it is "possible they are increasing their capability.... Iranians have also been known to boast and exaggerate their statements about greater technical and tactical capabilities."[191]

In one account, Iran reportedly claimed to have launched an improved "intermediate-range ballistic missile likely a Shahab-3 missile,"[192] but in another report a Pentagon official said Iran tested a Shahab-2 SRBM.[193]

Great Prophet 2 (November 2006)

Iran conducted military exercises throughout the country November 2-9, 2006. According to Iranian military leaders, "the first and main goal of this exercise is to demonstrate power and national determination to defend the country against any possible threat." [194]

Iran military officials added that other objectives included enhancing and testing the IRGC's deterrence capability and effectiveness; testing sophisticated and advanced weapons, conducting operational tactics involving distances of more than 1,400 kilometers; and, finally, coordinating land, naval, and air operations and logistics forces covering an area of 1,730 kilometers.[195] The Commander of the IRGC added that Iran was ready to share its missile systems with political allies and neighboring countries.[196]

Iran claimed to have test-fired "dozens of Shahab-2 and -3, Zolfaghar-73 [unknown what this referred to], Scud B, Fateh-110 and Zelzal" missiles in a central area of Iran.[197] It was the first time Iran claimed to have launched an enhanced range Shahab-3 MRBM during military maneuvers and said that it was fired with cluster warheads;[198] it was also reported to have an extended range of about 1,900 kilometers.[199] Iran further claimed to have launched three new types of tactical naval warfare missiles named Noor, Kowsar and Nasr, "suitable for covering all the Strait of Hormuz, the Persian Gulf and the Sea of Oman."[200] Vice Admiral Walsh, commander of U.S. naval forces in the region, expressed "particular concern over the threat the missiles pose to vessels traveling through the Sea of Oman and the Strait of Hormuz."[201]

Great Prophet 3 (July 2008)

Iran conducted a series of missile tests during military exercises in the Strait of Hormuz and in the desert from July 8-10, 2008. The missiles were launched from the Semnan Missile Test Range with warheads impacting within Iran or its coastal waters.[202] Although there was some confusion over what precisely occurred, U.S. officials and others determined that no new capabilities were demonstrated.[203] Iran reportedly launched some number of Shahab-1 and Shahab-2 SRBMs, a Shahab-3 MRBM, and Fateh, Zelzal, Nazeat, Fajr-3 and Oghab tactical missiles, but there is uncertainty over exactly how many. Iran and most early media accounts said nine missiles of varying ranges were launched.[204] But another report said that U.S. tracking systems detected only seven missile launches.[205] Television footage showed at least six missiles firing simultaneously, including the Shahab-3.[206] The following day, U.S. officials reported that one of the missiles had apparently failed the day before and that although Iran reported a second round of missile launches it appeared there was only one missile launched and it was unclear whether it was the one that had failed the day before.[207] Adding to the confusion were various reports concluding Iran had doctored or "photo-shopped" one of the photographs to make it appear that one of the missiles fired simultaneously with the others had launched successfully, instead of actually failing.[208] And a U.S. intelligence official was quoted as saying "while I cannot comment one way or another on these particular photos, it would be wrong to assume that the U.S. intelligence community accepts at face value what the Iranians disclose about their missile tests."[209] The State Department responded to the debate over how many launches actually occurred by noting that it did not matter precisely how many, but rather that Iran was testing increasingly longer and more capable missiles in the region.[210]

Great Prophet 4 (September 2009)

Iran conducted military and missile exercises from September 27 - 28, 2009 in which it reportedly launched the Shahab-3 MRBM and the new solid-fueled Sejil-2 MRBM.[211] An Iranian military official, Abdullah Araqi, was quoted as saying "Iranian missiles are able to target any place that threatens Iran."[212] Another source said the "optimized Shahab-3" missile has a range of 1,300 – 2,000 kilometers and that the Sejil was launched for the first time in military maneuvers from the central province of Semnan where Iran's space program is located.[213] The IRGC's Air Force Commander said the main aim of the exercise was to evaluate the "technical developments recently achieved in surface-to-surface missiles ... including simultaneous ... and successive" missile launches.[214] Also reported were launches specifically of Shahab-1 and 2 SRBMs, and Fateh, Tondar, Zelzal, and various other tactical ballistic missiles.[215] Iran claimed to have also tested a "multiple missile launcher for the first time."[216]

Great Prophet 5 (April 2010)

Iran conducted war games on April 22 – 25, 2010, reportedly firing five tactical sea-to-sea and shore-to-sea missiles at a single target simultaneously from different locations.[217] The Deputy Head of Iran's Armed Forces Headquarters, BG Massoud Jazayeri, said Iran "is designing defense operations to strengthen deterrent power of its forces and give a crushing response to any aggression."[218] But it does not appear that any short or medium-range ballistic missiles were launched in this exercise.

Great Prophet 6 (June 2011)

Iran conducted military exercises June 27-July 6, 2011 and reportedly launched 14 ballistic missiles, including Shahab-1 and Shahab-2 SRBMs and Zelzal and Fateh-110 tactical ballistic missiles. At least one Qiam SRBM, or an upgraded Shahab-3 or Ghadr (or Kadr), or a Sejil MRBM[219] was reportedly launched, depending on which source is cited. Reports said some missiles were aimed at targets at sea and perhaps as many as nine of the missiles were fired simultaneously.[220]

The Department of Defense subsequently confirmed that this exercise included a multiple missile salvo of some unspecified number.[221] Iran also unveiled the existence of a network of underground missile silos for the first time, which IRGC spokesman Asghar Qelich-Khani said were "part of the swift reaction unit of [Iran's] missile brigade; missiles are stored vertically."[222] The *New York Times* added that Iranian officials showed an underground launching pad or silo for what they called the Shahab-3 MRBM.[223] The televised reports also showed a large metal roof opening atop the silo to allow the firing of the missile.[224] Iran claimed the silos were built indigenously, but Israel was reported as saying the silos were built with DPRK assistance.[225]

One account of a tour of the silo complex also showed reported footage of a missile launch from a silo,[226] which at least one analyst said looked like the launch of a DPRK missile from its silo.

Although Iran kept the location of the silos a secret, others, such as *Jane's Defence Weekly*, said Iran's hidden silos were near Tabriz and Khorramabad.[227] IRGC deputy commander BG Hossein Salami further added "our missiles have aggressive, tactical, strategically deterrent and defensive features, of course we will not initiate any operation but our responses will be purely aggressive."[228]

The *Washington Post* reported a Shahab-3 was fired at targets at sea and added that Iranian television quoted Iranian military officials saying that Iran "began building a network of such silos across the nation 15 years ago."[229]

Great Prophet 7 (July 2012)

Iran held military exercises and missile launches July 1-3, 2012. Iran reported that it fired from different locations tens of Shahab-1 and Shahab-2 SRBMs, Shahab-3 MRBMs and Fateh, Qiam, Persian Gulf,[230] and Zelzal tactical ballistic missiles simultaneously at a mock air base in Iran's Lut desert in southeastern Iran.[231] IRGC Aerospace Force Commander BG Amir Ali Hajizadeh said "these maneuvers send a message to the adventurous nations that the IRGC is standing up to bullies alongside the determined and unified Iranian nation, and will decisively respond to any trouble they cause." Iran's FARS news agency further said these exercises "underline Tehran's threat to strike U.S. military bases in the neighboring countries—in Afghanistan, Bahrain, Kuwait and Saudi Arabia—if it comes under attack by Israel or the United States."[232] The *Associated Press (AP)* reported from Iranian sources that the missile tests demonstrated improved accuracy where 90% of the missiles hit their targets. Additionally, *AP* reported that Iran said it fired a considerable number of missiles against a single target (mock air base) making it "impossible for anti-missile systems to intercept and destroy them." Iran warned that 35 U.S. military bases in the Middle East are within Iran's missile range and would "be destroyed within seconds after any attack on Iran."[233] Another source said Iran claimed it had launched a missile capable of reaching targets 2,000 km away, but only fired to a range of 1,300 km. and repeated many of the things said here of July 2012 test launches.[234] There was no reported official U.S. response as to the accuracy of the claims made by Iran about their Great Prophet 7 exercises.

Appendix B. Reporting Requirements Relevant to Iranian Ballistic Missile Programs[235]

Title	721 Report		
Requirement	To identify acquisition by foreign countries over preceding six months of dual-use and other technology related to WMD and advanced conventional munitions, and trends of acquisition.		
Legislation	From	To	Frequency/ Duration
Sec. 721, Combating Proliferation of Weapons of Mass Destruction Act of 1996 (P.L. 104-293; 50 U.S.C. 2301 note)	Director of Central Intelligence	Intelligence Committees Speaker and minority leader of the House, majority and minority leaders of the Senate	Annually, by February 1
Title	ACDA Compliance Report		
Requirement	To identify: arms control, nonproliferation, disarmament objectives for upcoming year, including assessment of ongoing negotiations, U.S. adherence to its obligations, other nations' adherence, and noncompliance by foreign governments.		
Legislation	From	To	Frequency/ Duration
Sec. 403, Arms Control and Disarmament Act (P.L. 87-297; 22 U.S.C. 2593a	President (prepared by Secretary of State, with concurrence of Director of Central Intelligence, in consultation with Secretaries of Defense and Energy and Chairman of Joint Chiefs of Staff)	Speaker of the House, chairman of Senate Foreign Relations Committee	Annually, not later than April 15

Title	Annual Report on Proliferation of Missiles and Weapons of Mass Destruction		
Requirement	To identify information on the transfer in the preceding calendar year, by most countries, of "weapons, technology, components, or materials that can be used to deliver, manufacture, … or weaponize nuclear, biological, chemical or radiological weapons…."		
Legislation	From	To	Frequency/ Duration
Sec. 1308, Security Assistance Act of 2002 (P.L. 107-228; 50 U.S.C. 2368)	President	Committees on Appropriations, Armed Services, Foreign Affairs of the House; Committees on Appropriations, Armed Services, Foreign Relations of the Senate	Annually, not later than March 1
Title	Annual Threat Report		
Requirement	To identify: threats posed to the United States and its allies by weapons of mass destruction, ballistic missiles, and cruise missiles, and by their proliferation; states and non-state entities that have WMD and delivery systems, have related capability, or seek to develop such means and likely timelines for the development of the threat; relevant doctrines and control mechanisms.		
Legislation	From	To	Frequency/ Duration
Sec. 234, National Defense Authorization Act for Fiscal Year 1998 (P.L. 105-85; 50 U.S.C.2367)	Secretary of Defense (in consultation with Director of Central Intelligence)	Congress	Annually, by January 30
Title	Illicit Trade With Iran		
Requirement	To identify countries suspected of diverting certain goods, services, and technologies to or through Iran, including material contributions to Iran's development of nuclear, chemical, or biological weapons; ballistic missile or advanced conventional weapons capabilities; or support for international terrorism; and are Commerce Control List (CCL) items, United States Munitions List (USML) items, or prohibited under U.N. Security Council requirements.		
Legislation	From	To	Frequency/ Duration
Sec. 302, Comprehensive Iran Sanctions, Accountability, and Divestment Act of 2010 (P.L. 111-195; 22 U.S.C. 8542)	Director of National Intelligence, who reports within the Executive Branch to the President and Secretaries of Defense, State, Commerce, and Treasury	Committees on Finance; Banking, Housing, and Urban Affairs; and Foreign Relations of the Senate; and Committees on Ways and Means; Financial Services; and Foreign Affairs of the House	Annually, and when new information becomes available
Title	Iraq Sanctions Act Report (later made to also apply to Iran)		
Requirement	To identify sanctions taken by other nations against Iraq and Iran.		
Legislation	From	To	Frequency/ Duration
Sec. 586J(c), Iraq Sanctions Act of 1990 (P.L. 101-513; 50 U.S.C. 1701 note)	President	Committees on Appropriations and Foreign Relations of the Senate; Committees on Appropriations and Foreign Affairs of the House	Semiannually, with first report not later than May 5, 1991

(Continued)

Title	Non-State Entities Report		
Requirement	To identify: nuclear weapons and delivery system programs and related programs of nonnuclear-weapons states; nuclear weapons aspirations of non-state entities; and foreign persons who make a material contribution to such aspirations.		
Legislation	From	To	Frequency/ Duration
Sec. 1055, National Defense Authorization Act for Fiscal Year 2010 (P.L. 111-84; 50 U.S.C. 2371)	Director of National Intelligence	Committee on Armed Services and Select Committee on Intelligence of the Senate; Committee on Armed Services and Permanent Select Committee on Intelligence of the House	Biennially, with first report not later than September 1, 2010
Title	Proliferation Relating to Iran, North Korea, and Syria (INKSA Report)		
Requirement	To identify every foreign person for whom there is credible information that the person has transferred to or acquired from Iran, Syria, or North Korea goods, services or technology controlled for WMD and missile-related proliferation concerns.		
Legislation	From	To	Frequency/ Duration
Sec. 2, Iran, North Korea, and Syria Nonproliferation Act (P.L. 106-178; 50 U.S.C. 1701 note)	President	Committee on Foreign Affairs of the House; Committee on Foreign Relations of the Senate	Semiannually, with the first report not later than September 14, 2000
Title	Russian Proliferation to Iran and Other Countries		
Requirement	To identify Russian proliferation of WMD and ballistic missile goods, technology, expertise, and information, and of dual-use items to Iran and to other countries identified by the Director of Central Intelligence as a proliferation concern during the year preceding the year.		
Legislation	From	To	Frequency/ Duration
Sec. 1206, Bob Stump National Defense Authorization Act for Fiscal Year 2003 (P.L. 107-314; 22 U.S.C. 5952 note)	President	Congress	Not later than March 15 of 2003 through 2009
Title	WMD Threat		
Requirement	To assess the current threat of attack on the United States using ballistic missiles or cruise missiles; or chemical, biological, or nuclear weapon delivered by a system other than a ballistic missile or cruise missile		
Legislation	From	To	Frequency/Duration
Sec. 114, National Security Act of 1947 (P.L. 80-253; 50 U.S.C. 404i)	Director of National Intelligence	Committees on Intelligence, Foreign Relations, and Armed Services of the Senate; Committees on Intelligence, Foreign Affairs, and Armed Services of the House	First report required not later than December 31, 1991; after that annually by February 1

Source: United States Code.

Appendix C. Other Estimates of Iran's Ballistic Missiles

Table C-1. Iranian SRBMs (Other Sources)

	2000	2001	2002	2003	2004	2005	2006	2007	2008	2009	2010	2011	2012
	Jane's Strategic Weapon Systems												
Shahab 1 and Shahab 2 (Scud B and Scud C)			some number have been acquired, assembled, produced and used to date				300 - 400 missiles			around 50 launchers; 200 - 300 missiles	around 50 launchers; 200 - 300 missiles	around 50 launchers; 200 - 300 missiles	around 50 launchers; 200 - 300 missiles
Tondar 69 (CSS- 8)			some number exported from China in 1992							up to 30 launchers; up to 200 missiles	around 20 launchers; less than 100 missiles	around 20 launchers; less than 100 missiles	around 20 launchers; 200 missiles
Fateh A-110		tested	low-rate production		initial operational capability					3 versions may be in service	3 versions may be in service		
The Military Balance (The International Institute for Strategic Studies)													
Scud B / Scud C Shahab-1/2	est. 10 launchers; 300 missiles	est. 17 launchers; 300 missiles	est. 17 launchers; 300 missiles	est. 17 launchers; 300 missiles	12-18 launchers; 300 missiles	12-18 launchers; 300 missiles	12-18 launchers; 300 missiles	12-18 launchers; 300 missiles	12-18 launchers; 300 missiles	12-18 launchers; 300 missiles	12-18 launchers; 300 missiles	12-18 launchers; 300 missiles	12-18 launchers; 300 missiles
CSS-8	est. 25 launchers; 150 missiles	est. 30 launchers; 175 missiles	est. 30 launchers; 175 missiles	est. 30 launchers; 175 missiles	est. 30 launchers; 175 missiles	est. 30 launchers; 175 missiles	est. 30 launchers; 175 missiles	est. 30 launchers; 175 missiles	Est. 30 launchers; 175 missiles	Est. 30 launchers; 175 missiles	Est. 30 launchers; 175 missiles	Est. 30 launchers; 175 missiles	Est. 30 launchers; 175 missiles
Shaheen-1 Hatf-4 / Shaheen-2						some	some	some	some	some		some	some

Source: As noted. Missile designations as used by the source.

Notes: Where launcher numbers only are noted, actual missile inventory may be larger because launchers can be reused to fire additional missiles.

Table C-2. Iranian MRBMs (Other Sources)

	2000	2001	2002	2003	2004	2005	2006	2007	2008	2009	2010	2011	2012
Jane's Strategic Weapon Systems													
Shahab 3 and variants			production underway	production underway					< 20 TELs				25
Sejil / Ashura									testing	testing		IOC	
Shahab- 3/4 (Ghadr-1)				IOC	~20		30-50						
The Military Balance (The International Institute for Strategic Studies)													
Shahab-3	some	some (20 missiles)	some	at least 3	at least 6 launchers (Shahab 3 / Zelzal-3)	6 launchers; each with est. 4 missiles	est. 6 launchers; each with est. 4 missiles	est. 6 launchers; each with est. 4 missiles	est. 6 launchers; each with est. 4 missiles	est. 6 launchers; each with est. 4 missiles	up to 12 launchers, some Ghadr (or Kadr)	6 launchers, some Ghadr	12+ Shahab 3/ Ghadr
Sajjil													
Sajjil-2												In development	some (in development)

Source: As noted. Missile designations as used by the source.

Notes: Where launcher numbers only are noted, actual missile inventory may be larger because launchers can be reused to fire additional missiles.

End Notes

[1] After an initial powered phase of flight, a ballistic missile leaves the atmosphere and follows an unpowered trajectory or flight path before reentering the atmosphere toward a predetermined target. Ballistic missile ranges can vary from a hundred or so kilometers to more than 10,000 kilometers. The flight path may not be optimal if the trajectory is depressed or lofted relative to the optimal range trajectory. Differences in trajectories also depend on the type of missile (i.e., solid or liquid-fuel) and other factors. A space launch generally differs from a ballistic missile in that the launcher will fly a steeper trajectory and the burn time will be longer to get the satellite into orbit. At about 750 km from the launch site, these differences start to become apparent, thus allowing an assessment to be made about the purpose of a long-range ballistic missile test. There are also other factors that would allow determination of whether a launch is aimed at placing a satellite into orbit or delivering a warhead. Perhaps the quickest is to determine the direction of the launch vehicle. In Iran's case, a satellite launch would be in a south easterly direction to avoid overflight of other countries and to take advantage of the rotation of the earth.

[2] The Pentagon and the Missile Defense Agency (MDA) have long argued the value of BMD to provide for extended deterrence, devalue missile proliferation, dissuade ballistic missile investment, and enable international cooperation for regional defense.

[3] These would include the three-stage weapon currently deployed at the existing GMD (Ground-based Midcourse Defense) sites in Alaska and California, a two-stage version of the GMD missile first envisioned for deployment in Poland during the Bush Administration, and several versions of the Navy's SM-3 (Standard Missile-3).

[4] CRS Report RL32048, *Iran: U.S. Concerns and Policy Responses*, by Kenneth Katzman.

[5] Ibid. Generally speaking, one of Iran's primary objectives is to compel the United States and other outside powers to vacate the Gulf so that Tehran can exercise greater hegemony. Iran maintains the Gulf States, if left alone, could live in peace and security. The United States and many others disagree with this.

[6] The White House, *National Security Strategy*, Washington, DC, May 2010, p. 26.

[7] CRS Report RL32048, *Iran: U.S. Concerns and Policy Responses*, by Kenneth Katzman.

[8] Department of Defense, DOD News Briefing with Press Secretary Geoff Morrell from the Pentagon, February 3, 2009, http://www.defense.gov/transcripts/transcript.aspx? transcrip tid =4347.

[9] "Iran: Russia Says New Rocket Raises Nuclear 'Suspicions,'" *Associated Press*, February 7, 2008.

[10] James. R. Clapper, *Unclassified Statement for the Record on the Worldwide Threat Assessment of the U.S. Intelligence Community for the Senate Select Committee on Intelligence*, Director of National Intelligence, Washington, DC, January 31, 2012, p. 6.

[11] Pentagon, *Ballistic Missile Defense Review Report*, Washington, DC, 2010, p. 5.

[12] See footnote 11.

[13] See, for instance, Nick Braveman, "Iranian Nukes are not Heading Toward Tel Aviv," *Military Technology*, vol. 36, no. 2 (February 2012), p. 5.

[14] Paul Rogers, *Iran: Consequences of a War*, Oxford Research Group, Briefing Paper, Oxford, England, February 2006, p. 10.

[15] CRS Report R42678, *Conventional Arms Transfers to Developing Nations, 2004-2011*, by Richard F. Grimmett and Paul K. Kerr, p. 14.

[16] National Intelligence Council, *Foreign Missile Developments and the Ballistic Missile Threat to the United States Through 2015*, Washington, DC, September 1999. The most recent public update is National Intelligence Council, *Foreign Missile Development and the Ballistic Missile Threat Through 2015*, Washington, DC, December 2001.

[17] "Iran's Military Power," Lieutenant General Ronald L. Burgess, Jr., United States Army, Director, Defense Intelligence Agency, Statement before the Committee on Armed Services, U.S. Senate, April 14, 2010.

[18] Annual Report on Military Power of Iran, Department of Defense, April 2012, p. 4.

[19] Annual Report on Military Power of Iran, Department of Defense, April 2012, p. 1.

[20] See CRS Report RL34051, *Long-Range Ballistic Missile Defense in Europe*, by Steven A. Hildreth and Carl Ek. CRS Report R42529, *NATO's Chicago Summit*, by Paul Belkin. Also, LTG Patrick J. O'Reilly, USA, *BMD Overview— Phased Adaptive Approach*, Missile Defense Agency, 11-MDA-6321, Washington, DC, August 15, 2011, http://www.mda.mil /global/documents/pdf/bmds_briefing12.pdf.

[21] Previously, at the November 2010 Lisbon Summit, NATO formally agreed to implement a territorial BMD capability designed to cover all Member states in Europe, as well as the United States and Canada.

[22] CRS Report R42529, *NATO's Chicago Summit*, by Paul Belkin.

[23] Executive Summary of the Report of the Commission to Assess the Ballistic Missile Threat to the United States, July 15, 1998, pp. 21-22.

[24] For an assessment of Iran's nuclear program see CRS Report RL34544, *Iran's Nuclear Program: Status*, by Paul K. Kerr.

[25] Gen. Eugene Habiger, former Commander US Strategic Command, has said that "the miniaturization of a nuclear warhead is probably the most significant challenge any proliferant would have to face." See General Eugene Habiger – Carnegie Endowment for International Peace, September 20, 2004. Problems and Prospects of New Alaska Missile Interceptor Site. http://carnegieendowment.org/files/Habiger.pdf. See also Joe Cirincione and Elise Connor, How Iran Can Build a Bomb, Ploughshares Fund, July 2, 2010, http://www.ploughshares.org/blog/2010-07-02/how-iran-canbuild-bomb.

[26] The Annex Handbook of the Missile Technology Control Regime (MTCR) contains among other things considerable information about solid rocket propellants and their complexity. Additionally, a classic textbook for aerospace engineers similarly demonstrates the complexity associated with formulating and implementing a propellant for a given rocket motor. See George P. Sutton and Donald M. Ross, "Solid Propellant Rocket Fundamentals," in *Rocket Propulsion Elements: An Introduction to the Engineering of Rockets*, 5th ed. (New York, NY: John Wiley & Sons, Inc., 1986).

[27] Michael Elleman, *Iran's Ballistic Missile Capabilities: A Net Assessment* (London: The International Institute for Strategic Studies, 2010), p. 100.

[28] E-mail correspondence with author, September 12, 2012.

[29] Aaron Karp was one of the first to write of this in *Ballistic Missile Proliferation: The Politics and Technics* (London: SIPRI/Oxford University Press, 1996), pp. 51-98.

[30] U.S. Congress, Senate Committee on Armed Services, *Threat Assessment; Military Strategy; and Operational Requirements*, 101st Cong., 2nd sess., December 12, 1989; January 23-26, 30; February 2, 6-8, 21-22; March 7, 1990, S.Hrg. 101-780 (Washington: GPO, 1990), Statement of Hon. William H. Webster, Dir., Central Intelligence, p. 61.

[31] *Ballistic and Cruise Missile Threat*, National Air Intelligence Center, Wright-Patterson Air Force Base, Ohio, September 2000. The 2000 report identified several countries that were developing medium-range ballistic missiles but none were of that range and none had been deployed. These included North Korea (1,300-2,000 km.), India (2,000- 3,200 km.), Pakistan (1,300-1,900 km.), and Iran (1,300-2,400 km.). Two other countries, generally not considered Third World, had some number of medium-range ballistic missiles with max ranges of 2,400 km (Saudi Arabia, which received some number of Chinese CSS-2 missiles) and 2,800 km. (China). North Korea's Taepo-Dong 2, estimated by some with a max range of more than 5,470 km was the one Third World country considered developing a long-range ballistic missile at the time.

[32] For instance, in 2002, the Department of Defense assessed that China would deploy the DF-31 by mid-decade. This did not occur, although DOD assessed the DF-31 in 2007 as having an "initial threat availability" and assessed that less than 10 were deployed in 2008; that number has not changed. DOD also assessed in 2002 that the follow-on extended ICBM version of the DF-31 (DF-31A) would be deployed by mid-to-late decade, which occurred in 2008. In 2003, DOD correctly assessed that China could have up to 60 ICBMs by the end of the decade. Similarly, the Chinese SLBM program shows such problems. Since 2000, NASIC has assessed that China may be developing a medium-range submarine-launched ballistic missile, but NASIC never considered it actually deployed. In contrast, the Department of Defense reported that China had one XIA class submarine for those missiles (referred to variously as the JL-1 or CSSNX-3) as early as 2001. In 2002, DOD reassessed that China would deploy the JL-1 before the end of the year and moved its estimated date for deployment into 2003 the following year. DOD then stated from 2005-2007 that the XIA had deployed SLBMs, but then in 2008 and 2009 questioned whether the submarine was operational. In its latest report (2010), DOD questioned whether either the submarine or its associated ballistic missiles were operational. In 2002, the Department of Defense assessed that China would deploy the follow-on SLBM version of the DF-31 (JL-2 SLBM) by mid-to-late decade, which DOD indicated occurred in 2008-2009. But, in 2010, DOD assessed the JL-2 SLBM as still developmental. A flight test failure of the JL-2 SLBM in summer 2004 reportedly delayed its initial deployment[32] and the 2010 DOD report said the JL-2 failed several of its final rounds of tests. DOD currently believes the JL-2 SLBM and associated submarine combination renders uncertain when it will become operational at all.

[33] Initial growth estimates made by the Department of Defense in 2002 were that China would add about 50 SRBMs each year, but by 2003 DOD revised this estimate upward to over 75 missiles each year and in 2006 increased this further to about 100 per year. This trend appears to have slowed considerably or even halted in 2009-2010.

[34] This is taken from congressional testimony by the author, "Some Thoughts on ICBM Development and ICBM Threats." See U.S. Congress, House Committee on Oversight and Government Reform, Subcommittee on National Security and Foreign Affairs, *Oversight of Ballistic Missile Defense (Part 1): Threats, Realities and Tradeoffs*, 110th Cong., 2nd sess., March 5, 2008, Serial No. 110–148 (Washington: GPO, 2009), pp. 55-72.

[35] Technically, the UK did not develop their strategic missiles; Trident II is purchased from the United States.

[36] One might argue that India did not need an ICBM 30 years ago, rather than striving for that long to develop and test an ICBM.

[37] "Iran's Guard Commander comments on Tehran's missile power," *BBC Monitoring Middle East - Political*, November 13, 2006, source: *Vision of the Islamic Republic of Iran Network 2*, Tehran, in Persian, November 12 2006.

[38] William Branigin, Thomas Erdbrink, and Walter Pincus, "Iran Test Fires Its Most Advanced Missiles," *The Washington Post*, September 28, 2009, On-line edition.

[39] "Iran doctored missile test-firing photo: defence analyst," *Agence France Presse - English*, July 9, 2008, and Al Kamen, "Iran Apparently in Possession of Photoshop," *The Washington Post*, July 11, 2008, p. A15, Suburban Edition.

[40] "Iran test-fires 'long-range' naval missile," *Agence France Press - English*, August 27, 2006, newswire. Iranian Navy Commander Admiral Koochaki was quoted as saying "the highly destructive Sagheb was successfully shot from a submarine but it can be carried on any other (naval) vessel."

[41] "World in Brief," *The Washington Post*, September 10, 2006, p. A23. Pentagon officials confirmed the test was "bogus" and U.S. intelligence determined the video matched a video of an earlier Chinese test.

[42] Siavosh Ghazi, "Iran launches home-built satellite rocket," *Agence France Presse - English*, August 17, 2008, newswire. Both the head of Iran's space agency and Iran's Defense Ministry said the Safir rocket carrying the Omid satellite launched successfully. Iranian officials later admitted that Iranian state media mistakenly claimed the satellite entered orbit.

[43] Nazila Fathi and Thom Shanker, "Iran sees itself assisting allies with satellites," *The New York Times*, August 19, 2008, p. A11. The article cited Pentagon and U.S. military officials who said the first stage of the missile performed successfully, but the second stage failed causing the missile to fly off wildly and destroying the top of the missile and its nose cone. Another source said advance notice of the test allowed the USS Russell, an Arleigh Burke-class destroyer operating in the Persian Gulf to monitor and radar track the launch. Infrared data from the U.S. Air Force Defense Support Program (DSP) missile warning satellite provided additional intelligence. Craig Covault, "U.S. radar and satellite data indicate Iran launch failed," *Aerospace Daily & Defense Report*, vol. 227, no. 6 2008, p. 1.

[44] Dennis C. Blair, Dir., National Intelligence, *Annual Threat Assessment of the U.S. Intelligence Community for the Senate Select Committee on Intelligence*, February 2, 2010, p. 25.

[45] Lt. Gen. Ronald L. Burgess, Jr., Dir., Defense Intelligence Agency, *Iran's Military Power, Statement before the Senate Committee on Armed Services*, April 14, 2010, p. 1.

[46] Dennis C. Blair, Dir., National Intelligence, *Annual Threat Assessment of the U.S. Intelligence Community for the Senate Select Committee on Intelligence*, February 2, 2010, pp. 13-14.

[47] Lt. Gen. Ronald L. Burgess, Jr., Dir., Defense Intelligence Agency, *Iran's Military Power, Statement before the Senate Committee on Armed Services*, April 14, 2010, p. 13.

[48] Anthony H. Cordesman and Aram Nerguizian, *The Gulf Military Balance in 2010: An Overview*, Center for Strategic & International Studies, Washington, DC, April 2010, p. 36. Data for military expenditures was taken from the annual Military Balance, published by the International Institute for Strategic Studies (London).

[49] Ibid., Figure 24: Southern Gulf Military Expenditures by Country: 1997-2009, p. 36.

[50] Ibid., Figure 25: Comparative Military Expenditures of the Gulf Powers as a Percent of GDP – 1989-2009.

[51] The World Fact book, CIA: https://www.cia.gov/library/publications/the-world-factbook /index.html

[52] Some of this information has been gathered over the years from what little information is published in the *Federal Record* regarding U.S. sanctions against specific Iranian entities. Additionally, UN sanctions reports and other UN reports can also provide information about Iranian businesses and other entities involved with Iran's ballistic missile and space launch infrastructure.

[53] Orbital launches are conducted from Semnan, though some sub-orbital sounding rocket launches may have taken place elsewhere.

[54] The Nuclear Threat Initiative (www.nti.org) maintains a highly useful accounting of these and a wide variety of other facilities and suspect sites not just in Iran, but elsewhere too. Additionally, Iran Watch (part of the Wisconsin Project on Arms Control) maintains a large database of suspect ballistic missile facilities in Iran (www.iranwatch.org).

[55] Kenneth Katzman, *The Warriors of Islam: Iran's Revolutionary Guard* (Boulder, CO: Westview Press, 1993).

[56] Lt. Gen. Ronald L. Burgess, Jr., Dir., Defense Intelligence Agency, *Iran's Military Power, Statement before the Senate Committee on Armed Services*, April 14, 2010, p. 13.

[57] Michael Elleman, *Iran's Ballistic Missile Capabilities: A Net Assessment* (London: The International Institute for Strategic Studies, 2010), p. 13.

[58] Ibid. Elleman points out, however, that open-source literature provides conflicting accounts of Iran's indigenous capabilities in this regard. Some analysts argue Iran has made significant strides in their own ability to produce the Shahab-1 and Shahab-2 missiles, and others argue that Iran must still import key components if not entire missiles from countries such as North Korea. See Elleman, *Iran's Ballistic Missile Capabilities*, p. 36.

[59] Michael Elleman, *Iran's Ballistic Missile Capabilities: A Net Assessment* (London: The International Institute for Strategic Studies, 2010), p. 32.

[60] United Nations, Security Council Committee established pursuant to resolution 1737 (2006, *Final Report of the Panel of Experts submitted in accordance with resolution 1984 (2011)*, S/2012/395, June 12, 2012. The Panel of Experts report was issued under UN Security Council Resolution 1929, which mandated the Panel of Experts. UN Security Council Resolution 1984 renewed the 1929 Panel of Experts mandate.

[61] *Iran's Guards Get Qiam Missiles, Israel Claims are Bound to Hezbollah,* Al-Manar TV Lebanon, May 23, 2011.

[62] Katz, Yaakov, *Israel; New Missile Like Bound for Hezbollah*, Jerusalem Post, May 23, 2011.

[63] "Iran test fires short-range missile," *Aljazeera*, August 4, 2012, online edition.

[64] Congress established that a bi-annual (now annual) report from the Director, Central Intelligence, be submitted to identify acquisition and trends by foreign countries technology related to WMD and advanced conventional weapons. The 721 Report derives its name from Section 721 (which requires this report) of the Combating Proliferation of Weapons of Mass Destruction Act of 1996. P.L. 104-293; 50 U.S.C. 2301 note.

[65] "Israel 'highly sceptical' of IAEA deal with Iran," *Space Daily*, May 22, 2012, on line.

[66] Lt. Gen. Ronald L. Burgess, Jr., Dir., Defense Intelligence Agency, *Iran's Military Power, Statement before the Senate Committee on Armed Services*, April 14, 2010, p. 13.

[67] BG Itai Brun, head of Israel's intelligence service research division, testified before Israel's parliamentary committee on foreign affairs and defence, that Iran "currently had 450 missiles capable of hitting the Jewish state. See "Iran's Guard Commander comments on Tehran's missile power," *BBC Monitoring Middle East - Political*, November 13, 2006, source: *Vision of the Islamic Republic of Iran Network 2*, Tehran, in Persian, November 12, 2006.

[68] Uzi Rubin, "Showcase of Missile Proliferation: Iran's Missile and Space Program," *Arms Control Today*, January/February 2012.

[69] Uzi Rubin, "The Global Range of Iran's Ballistic Missiles."

[70] Portions of Israel and other targets in the region remained out of reach of the Shahab-3. In addition, Iran apparently did not want to deploy its MRBMs on the borders of Iraq where it could be more vulnerable to attack than if moved further inland. Iranian efforts to develop a 2,000 km or more MRBM, therefore, perhaps in underground bunkers or silos further into the interior of Iran, seemingly are important drivers of Iran's program.

[71] Uzi Rubin, "The Global Reach of Iran's Ballistic Missile Program," *Jerusalem Issue Brief (Jerusalem Center for Public Affairs)*, vol. 5, no. 26 (June 20, 2006).

[72] This unreleased UN report is identified as: United Nations, *Final Report*, Panel of Experts Established Pursuant to resolution 1929 (2010), May 2010. The only online copy that could be found is: http://www.scribd.com /doc/55737041/ Leaked-UN-Panel-of-Experts-Report-on-Iran-Sanctions-May-2011.

[73] Michael Elleman, *Iran's Ballistic Missile Capabilities: A Net Assessment* (London: The International Institute for Strategic Studies, 2010), p. 20.

[74] The article further said "the Iranian warhead is designed to carry a spherical nuclear weapon that would be detonated 2,000 feet above the ground.... Although U.S. officials believe Iran is several years away from acquiring nuclear weapons, they point out that the warhead could hold a version of the nuclear bomb Pakistan is known to have developed. Iran has acquired a detailed breakdown of Pakistan's nuclear weapons." Con Coughlin, "Iran has missiles to carry nuclear warheads. Updated Scud puts Israel in Teheran's sights," *The Daily telegraph (London)*, April 7, 2006, p. 19.

[75] "Iran's Guard Commander comments on Tehran's missile power," *BBC Monitoring Middle East - Political*, November 13, 2006, source: *Vision of the Islamic Republic of Iran Network 2*, Tehran, in Persian, November 12, 2006.

[76] Michael Elleman, *Iran's Ballistic Missile Capabilities: A Net Assessment* (London: The International Institute for Strategic Studies, 2010), p. 110.

[77] Uzi Rubin, "The Global Reach of Iran's Ballistic Missile Program," *Jerusalem Issue Brief (Jerusalem Center for Public Affairs)*, vol. 5, no. 26 (June 20, 2006).

[78] These terms are generally used within industry. Usually, a knock-down kit includes all the parts for a system, but is shipped unassembled. Semi-knock-down can mean that some of the harder parts—engine installation, hydraulic plumbing, electronics wiring—have been put together in advance, with only the minimal-skill items left for assembly by the recipient.

[79] Robert H. Schmucker and Markus Schiller, "The DPRK Missile Show: A Comedy in (Currently) Eight Acts," Draft report, Schmucker Technologies, Report XX/10, Germany, May 5, 2010, p. 12-13. The authors cite media reports that North Korea had sold to Iran a reverse engineered and improved version of the Soviet R-27/SSN-6 now named BM-25. Schmucker's earlier assessment of Iran's missiles includes the modified R-27/SSN-6 as procured via North Korea without any DPRK modification, further describing its status as unknown and that there was no production of or modification of the missile in Iran. See Robert H. Schmucker, "Iran and its Regional Environment," Second Transatlantic Conferencc, Berlin, Germany, March 27, 2006, pp. 4, 37-39.

[80] John Pomfret and Walter Pincus, "Experts question North Korea-Iran missile link from WikiLeaks document release," *The Washington Post*, December 1, 2010, online edition.

[81] Michael Elleman, *Iran's Ballistic Missile Capabilities: A Net Assessment* (London: The International Institute for Strategic Studies, 2010), p. 32.

[82] John Pomfret and Walter Pincus, "Experts question North Korea-Iran missile link from WikiLeaks document release," *The Washington Post*, December 1, 2010, online edition.

[83] Theodore Postol, *A Technical Assessment of Iran's Ballistic Missile Program. Technical Addendum to the Joint Threat Assessment on Iran's Nuclear and Missile Potential*, EastWest Institute, Iran's Nuclear and Missile Potential: A Joint Threat Assessment by U.S. and Russian Technical Experts, by Siegfried Hecker and David Holloway, Stanford University, Stanford, CA, May 2009, pp. 20-24.

[84] "Iran Test Fires Missiles, Unveils Secret Silos," *The Frontrunner - English*, June 29, 2011, newswire.

[85] William J. Broad, "Iran shows off strengthened underground silos for missiles," *The New York Times*, June 29, 2011, p. 5.

[86] United Nations, Security Council Committee established pursuant to resolution 1737 (2006, *Final Report of the Panel of Experts submitted in accordance with resolution 1984 (2011)*, S/2012/395, June 12, 2012, p. 22.

[87] Various secondary sources cite Hiroshi Kano, "Tehran, Pyongyang Push Arms," *International Review*, Autumn 1994 for this account, but CRS was unable to find this article in any journal. After an exhaustive search of media and journal articles, CRS was able to find only one report at the time that said anything about such shelters. According to *The Jerusalem Post*, "sources say they believe Foruzandeh and North Korea discussed the possibility of joint production of the [Nodong], along with underground shelters for the missiles." See David Makovsky, in Israel said not pushing U.S. to stop N. Korean missile sales to Iran," *The Jerusalem Post*, December 24, 1993. Although the news account demonstrates a relatively weak case, the report is worth mentioning as it fits the general time frame for when Iran may have first begun its efforts to build such shelters, which were then under construction not long thereafter.

[88] Martin Sieff and Bill Gertz, "Iran's long-range missile plans worry Netanyahu; Says U.S. cities may be targeted," *The Washington Times*, October 1, 1998, p. A13.

[89] Uzi Rubin, "The Global Reach of Iran's Ballistic Missile Program," p. 65, http://jcpa.org/text/iran_page_62-67.pdf.

[90] Sean O'Connor, who runs gemint.com, told the author in March 2011 that he thought he did "the first open-source look at Iran's silos in February of 2008." This particular blog is no longer available on his website, but the copy he sent to the author said there had been "a great deal of recent open-source reporting dealing with a purported missile site near Tabriz in northwestern Iran. While these reports may or may not be true, they have all failed to note the presence of another missile facility in the region [about 10 kilometers southwest of Tabriz]. The difference is that this facility can be identified with much greater certainty: this facility houses Iran's first missile silos." A copy of this specific blog can be found elsewhere, however, at http://www.irandefence.net/showthread.php?t=29952. O'Connor did not indicate where the open-source reporting came from and the author was not able find any reports about Iranian silos during this time. O'Connor has blogged many times since about the missile silos at Tabriz and Khorramabad, providing details and satellite imagery. Jeffrey Lewis at armscontrolwonk.com wrote of a suspect missile facility near Tabriz in February 2008 as well, based on rough translations of a German newspaper *(Bild)* article that might have been published as early as 2005 but was apparently no longer available. Although blog contributors provided various and somewhat different translations, it appears the original source was the BND (German intelligence

agency), which said Iran had bought 18 unassembled mobile BM-25 rockets (range of 2,500 km), which Iran remodeled to obtain a 3,500 km range, and that "German intelligence had located a stationary starting system for the test of silo-supported rockets [as translated]." Lewis' blog too has written about the silos at Tabriz and Khorramabad at various times since.

[91] Sean O'Connor, "Iranian SSM Facilities," *I&A (Imagery and Analysis)*, vol. 1, no. 2 (March 20, 2011), p. 11, self-published, http://www.gemint.com.

[92] James Kirkup, "Iran accused of secret missile tests," *The Daily Telegraph (London)*, June 30, 2011, p. 22. UNSCR 1929 prohibits Iran from any activities related to ballistic missiles capable of delivering nuclear weapons, including launches using ballistic missile technology, and provides that UN Member States are to take all measures necessary to prevent the transfer of technology or technical assistance to Iran related to such activities. Hague said these tests were conducted from October 2010 to some point before his speech in June 2011.

[93] "Britain accuses Iran of secretly testing nuclear-capable missiles," *States News Services*, June 29, 2011, newswire.

[94] Ibid.

[95] David Stringer, "UK: Iran conducting secret ballistic missile tests," *The Associated Press*, June 29, 2011, newswire.

[96] This unreleased UN report is identified as: United Nations, *Final Report*, Panel of Experts Established Pursuant to resolution 1929 (2010), May 2010. The only online copy that could be found is available at http://www.scribd.com/doc/ 55737041/Leaked-UN-Panel-of-Experts-Report-on-Iran-Sanctions-May-2011.

[97] The report said, "The Panel was informed of [these two Sejil launches] by a Member State." Ibid., p. 26.

[98] "Iran says fires missiles to Indian Ocean for first time," *Reuters*, July 9, 2011, newswire.

[99] Ibid.

[100] "Iran claims to have tested missiles in Indian Ocean," *Jane's Intelligence Weekly*, July 11, 2011.

[101] Remarks by Ambassador Susan E. Rice, U.S. Permanent Representative to the United Nations, at a Security Council Briefing on Iran and Resolution 1737, New York, NY, June 23, 2011.

[102] United Nations, Security Council Committee established pursuant to resolution 1737 (2006, *Final Report of the Panel of Experts submitted in accordance with resolution 1984 (2011)*, S/2012/395, June 12, 2012.

[103] India recently successfully tested a long-range ballistic missile just under what is defined as ICBM range (around 5,000 km), but it will still be several years before that missile class is considered operational and it is further unknown whether it will carry a nuclear warhead.

[104] U.S. Congress, Senate Select Committee on Intelligence, *Current and Projected National Security Threats to the United States*, 106th Cong., 2nd sess., February 2, 2000, S.Hrg. 106-580 (Washington: GPO, 2000), p. 11.

[105] National Intelligence Council, "Foreign Missile Developments and the Ballistic Missile Threat to the United States through 2015," September 1999. This unclassified National Intelligence Estimate was provided in open testimony to the Senate Foreign Relations Committee on September 16, 1999, by Robert D. Walpole, National Intelligence Officer for Strategic and Nuclear Programs, Central Intelligence Agency.

[106] National Intelligence Council, "Foreign Ballistic Missile Threat Through 2015," December 2001, Unclassified Summary of a National Intelligence Estimate.

[107] Lt. Gen. Ronald L. Burgess, Jr., Dir., Defense Intelligence Agency, *Iran's Military Power, Statement before the Senate Committee on Armed Services*, April 14, 2010, pp. 13-14.

[108] See CRS Report RL34051, *Long-Range Ballistic Missile Defense in Europe*, by Steven A. Hildreth and Carl Ek, and CRS Report CRS Report RL32048, *Iran: U.S. Concerns and Policy Responses*, by Kenneth Katzman.

[109] See National Intelligence Council, "Foreign Missile Developments, 1999," and Robert D. Walpole, "The Ballistic Missile Threat to the United States," Statement Before the Senate Subcommittee on International Security, Proliferation and Federal Services, February 9, 2000.

[110] United Nations, Security Council Committee established pursuant to resolution 1737 (2006, *Final Report of the Panel of Experts submitted in accordance with resolution 1984 (2011)*, S/2012/395, June 12, 2012, p. 23.

[111] For instance, see Charles P. Vick, "North Korean, Iranian, and Pakistani Common Russian, Chinese Nuclear Weapons Heritage and Tests, What does it Reveal about the Missile Borne Warhead Development Status?" Part 1, March 20, 2007, at http://www.globalsecurity.org /wmd/world/dprk/nuke-warhead-dev1.htm.

[112] Global Security: Iran, Fifth Report of Session, 2007-08. Report together with formal minutes, oral and written evidence, House of Commons, Foreign Affairs Committee, Ordered by The House of Commons to be printed 20 February 2008. HC 142, Incorporating HC 496-i, ii, iii Session 2006-2007. Published on 2 March 2008, London, p. Ev.27.

[113] Einar Koch, "German Intelligence Service Sees Iran Upgrading Missiles to Reach Israel, Europe," *Bild (Hamburg)*, December 16, 2005, (Internet Version-WWW).

[114] Gareth Porter, "Russians Refuted U.S. Claim of Iranian Missile Threat to Europe," *IPSNews.net*, November 30, 2010, On-line edition.

[115] "Iran does not have technology to build ICBMs - Russian Chief of Staff (Part 2)," *Russia & CIS Military Newswire*, November 2, 2006. See also "Iran "not capable" of creating intercontinental missiles: Russia," *Agence France Presse - English*, November 2, 2006, newswire.

[116] David E. Hoffman, "In WikiLeaks: Russia's doubts about Iranian missiles," *Foreign Policy Journal—online blog*, November 29, 2010.

[117] Jeffrey Heller and Mark Heinrich, "Israel: Iran will have U.S.-range missile in 2-3 years," *Reuters*, February 22, 2012, U.S., online edition.

[118] United Nations, Security Council Committee established pursuant to resolution 1737 (2006, *Final Report of the Panel of Experts submitted in accordance with resolution 1984 (2011)*, S/2012/395, June 12, 2012, p. 20.

[119] "Unclassified Report on Military Power of Iran," Department of Defense, April 2010, p. 11.

[120] Ali Sheikholeslami, "Iran Say No Plan for Missile Able to Reach U.S.," *Bloomberg Business Week*, April 21, 2010, Online edition and "Iran has no Plan to Build Missiles Capable of Reaching U.S.," *Tehran Times*, April 22, 2010, Online edition.

[121] "Iran Rejects Report on Transfer of N. Korean Missiles to Iran," *Fars News Agency (Tehran, Iran)*, November 29, 2010, On-line edition.

[122] "Guard Chief: Iran can build longer-range missiles," *The Associated Press*, June 28, 2011, newswire.

[123] "Iran not to make longer-range missiles," *Iran Government News*, June 28, 2011.

[124] "Iran's new space center boosts rocket plan," *UPI.com*, August 3, 2012, online edition.

[125] The more easterly the launch allows the launcher to gain maximum speed from the rotation of the earth.

[126] In order of first orbital launch are: the Soviet Union, the United States, France, Japan, China, the United Kingdom, India, Israel and Iran. Russia and Ukraine are considered to have inherited their space launch capabilities as successor states to the former Soviet Union. The European Space Agency (ESA) is not included because it is a consortium of a number of European governments and commercial companies.

[127] Stephen Clark, "Iran puts Satellite into Earth Orbit," *Spaceflight Now*, February 3, 2009, http://spaceflightnow.com/ news/n0902/03iran/.

[128] Iranian Space Agency, *In the Name of God the Compassionate the Merciful*, OMD Satellite Launch Report, 46th Meeting of the Scientific and Technical Subcommittee of the Committee on the Peaceful Uses of Outer Space (COPUOS), 9th-20th February 2009, February 9, 2009.

[129] Department of State, "Iranian Launch of Satellite," press release, February 3, 2009.

[130] Department of Defense, DOD News Briefing with Press Secretary Geoff Morrell from the Pentagon, February 3, 2009, http://www.defense.gov/transcripts/transcript.aspx? transcript id=4347.

[131] Robert H. Schmucker and Markus Schiller, "The DPRK Missile Show: A Comedy in (Currently) Eight Acts," Draft report, Schmucker Technologies, Report XX/10, Germany, May 5, 2010, p. 8.

[132] *Defense Minister: New Space Projects to be Unveiled*, Islamic Republic News Agency (IRNA), January 30, 2011.

[133] *Iran President Unveils Four New Satellites*, Islamic Republic News Agency (IRNA), February 7, 2011.

[134] Others are the United States, the Soviet Union, China, Japan and the European Space Agency.

[135] "Iran plans manned space mission in 10 years," *Reuters*, August 21, 2008, newswire.

[136] "Iran to send first astronaut into space within 10 years," RIA Novosti, August 20, 2008.

[137] Prior to the 1979 Revolution, Iran had ties to Western countries such as the United States. Iran had some it inherited from under the rule of the Shah. These were older, short-range tactical and battlefield missiles it used in the Iran-Iraq War (1980-1988).

[138] Michael Elleman, *Iran's Ballistic Missile Capabilities: A Net Assessment*, The International Institute for Strategic Studies, London, England, May 2010, pp. 13-14.

[139] E-mail correspondence with Michael Elleman (IISS), September 2012.

[140] U.S. Congress, Senate Committee on Armed Services, *Threat Assessment, Military Strategy, and Defense Planning*, 102nd Cong., 2nd sess., January 22, 1992, S.Hrg. 102-755 (Washington: GPO, 1992), pp. 8-21.

[141] U.S. Congress, Senate Committee on Governmental Affairs, *Proliferation Threats of the 1990's*, 103rd Cong., 1st sess., February 24, 1993, S.Hrg. 103-208 (Washington: GPO, 1993), pp. 8-12

[142] U.S. Congress, Senate Committee on Governmental Affairs, *Proliferation Threats of the 1990's*, 103rd Cong., 1st sess., February 24, 1993, S.Hrg. 103-208 (Washington: GPO, 1993), p. 192.

[143] In response to hearing questions, the IC responded thusly some months later. See U.S. Congress, Senate Committee on Governmental Affairs, *Proliferation Threats of the 1990's*, 103rd Cong., 1st sess., February 24, 1993, S.Hrg. 103-208 (Washington: GPO, 1993), p. 172.

[144] Foreign Broadcast Information Service, *Proliferation Issues. Russian Federation Foreign Intelligence Service Report. A New Challenge After the Cold War: Proliferation of Weapons of Mass Destruction*, JPRS-TND-93-007, Washington, DC, March 5, 1993, p. 29-30.

[145] "Iran admits buying missiles from North Korea during Iraq war," *BBC Monitoring Asia Pacific—Political*, November 7, 2006, translated from *Nezavisimaya Gazeta*, November 3, 2006, p. 7.

[146] CRS Report R42678, *Conventional Arms Transfers to Developing Nations, 2004-2011*, by Richard F. Grimmett and Paul K. Kerr, p. 44.

[147] Gates also said tens of thousands of scientists and engineers are emigrating from the former Soviet Union, and "some have expertise applicable to special weapons and missiles... some may find a better market for their expertise in Third World countries trying to acquire or improve special weapons capabilities." U.S. Congress, House Committee on Armed Services, *Potential Threats to American Security in the Post-Cold War Era*, Hearings Before the Defense Policy Panel, 102nd Cong., 1st sess., December 10, 1991, H.A.S.C. No. 102-52 (Washington: GPO, 1992), p. 9.

[148] U.S. Congress, Senate Committee on Armed Services, *Threat Assessment, Military Strategy, and Defense Planning*, 102nd Cong., 2nd sess., January 22, 1992, S.Hrg. 102-755 (Washington: GPO, 1992), pp. 8-21.

[149] U.S. Congress, Senate Committee on Governmental Affairs, *Proliferation Threats of the 1990's*, 103rd Cong., 1st sess., February 24, 1993, S.Hrg. 103-208 (Washington: GPO, 1993), pp. 8-12.

[150] U.S. Congress, Senate Select Committee on Intelligence, *Current and Projected National Security Threats to the United States and its Interests Abroad*, 103rd Cong., 2nd sess., January 25, 1994, S.Hrg. 103-630 (Washington: GPO, 1994), pp. 4-78.

[151] Perhaps a key reason, according to Director, DIA, Lt. Gen. James R. Clapper was that Iran's military was rebuilding from its war with Iraq and that major economic constraints, such as hard currency shortages and a poor debt servicing record limited Tehran's ability to acquire weapon systems in the international marketplace. U.S. Congress, Senate Select Committee on Intelligence, *Worldwide Intelligence Review*, 104th Cong., 1st sess., January 10, 1995, S.Hrg. 104- 15 (Washington: GPO, 1995), p. 34.

[152] This report is referred to as the 721 Report because it is a congressionally mandated under section 721 of the Intelligence Authorization Act for Fiscal Year 1997 (P.L. 104-293). Initially, it was a biannually-required report to Congress, but legislation changed that to an annual reporting requirement starting in 2004.

[153] Director of Central Intelligence, *The Acquisition of Technology Relating to Weapons of Mass Destruction and Advanced Conventional Munitions, July - December 1996*, Washington, DC, June 1997.

[154] 1997 Annual Threat Assessment.

[155] Director of Central Intelligence, *Report of Proliferation-Related Acquisition in 1997*, Washington, DC.

[156] Ibid.

[157] 1998 Annual Threat Assessment.

[158] Director of Central Intelligence, *Unclassified Report to Congress, January – June 1998*, Washington, DC.

[159] 1999 Annual Threat Assessment.

[160] Director of Central Intelligence, *Unclassified Report to Congress on the Acquisition of Technology Relating to Weapons of Mass Destruction and Advanced Conventional Munitions, 1 January through 30 June 1999*, Washington, DC.

[161] Director of Central Intelligence, *Unclassified Report to Congress on the Acquisition of Technology Relating to Weapons of Mass Destruction and Advanced Conventional Munitions, 1 July through 31 December 1999*, Washington, DC.

[162] National Intelligence Council (NIC), *Foreign Missile Developments and the Ballistic Missile Threat to the United States Through 2015*, Washington, DC, September 1999

[163] Deputy Director of National Intelligence for Analysis, *Unclassified Report to Congress on the Acquisition of Technology Relating to Weapons of Mass Destruction and Advanced Conventional Munitions, 1 January through 31 December 2010*, Washington, DC.

[164] Director of National Intelligence, James R. Clapper. *Statement for the Record on the Worldwide Threat Assessment of the U.S. Intelligence Community for the Senate Armed Services*, March 10, 2011. Also, Director, Defense Intelligence Agency, Lt. Gen. Ronald L. Burgess. *World Wide Threat Assessment. Statement before the Committee on Armed Services*, March 10, 2011.

[165] Director of National Intelligence, James R. Clapper. *Unclassified Statement for the Record on the Worldwide Threat Assessment of the U.S. Intelligence Community for the Senate Select Committee on Intelligence*, January 31, 2012.

[166] Deputy Director of National Intelligence for Analysis, *Unclassified Report to Congress on the Acquisition of Technology Relating to Weapons of Mass Destruction and Advanced Conventional Munitions, 1 January through 31 December 2010*, Washington, DC.

[167] Deputy Director of National Intelligence for Analysis, *Unclassified Report to Congress on the Acquisition of Technology Relating to Weapons of Mass Destruction and Advanced Conventional Munitions, 1 January through 31 December 2011*, Washington, DC.

[168] Robert H. Schmucker and Markus Schiller, "The DPRK Missile Show: A Comedy in (Currently) Eight Acts," Draft report, Schmucker Technologies, Report XX/10, Germany, May 5, 2010.

[169] Michael Elleman, *Iran's Ballistic Missile Capabilities: A Net Assessment* (London: The International Institute for Strategic Studies, 2010), p. 94.

[170] Theodore Postol, *Technical Addendum to the Joint Assessment on Iran's Nuclear and Missile Potential, A Technical Assessment of Iran's Ballistic Missile Program*, May 6, 2009.

[171] CRS Report R42678, *Conventional Arms Transfers to Developing Nations, 2004-2011*, by Richard F. Grimmett and Paul K. Kerr, p. 44.

[172] U.S. Congress, Senate Committee on Armed Services, *Threat Assessment, Military Strategy, and Defense Planning*, 102nd Cong., 2nd sess., January 22, 1992, S.Hrg. 102-755 (Washington: GPO, 1992), pp. 8-21, and also U.S. Congress, Senate Committee on Governmental Affairs, *Proliferation Threats of the 1990's*, 103rd Cong., 1st sess., February 24, 1993, S.Hrg. 103-208 (Washington: GPO, 1993), pp. 8-12.

[173] Foreign Broadcast Information Service, *Proliferation Issues. Russian Federation Foreign Intelligence Service Report. A New Challenge After the Cold War: Proliferation of Weapons of Mass Destruction*, JPRS-TND-93-007, Washington, DC, March 5, 1993, p. 29.

[174] 1996 Annual Threat Assessment.

[175] Director of Central Intelligence, *The Acquisition of Technology Relating to Weapons of Mass Destruction and Advanced Conventional Munitions, July - December 1996*, Washington, DC, June 1997.

[176] Director of Central Intelligence, *Report of Proliferation-Related Acquisition in 1997*, Washington, DC.

[177] 1998 Annual Threat Assessment.

[178] Ibid. The IC testified that China had agreed to abide by the "guidelines and parameters" of the Missile Technology Control Regime (MTCR) and had committed not to transfer ground-to-ground MTCR-class missiles. But China did not appear to interpret its responsibilities under the MTCR guidelines as strictly as the United States and other MTCR members the IC said. By all indications China had taken itself out of the business of exporting complete ballistic missiles, however. This, the IC assessed, was an important step—one that has slowed the process of military destabilization in South Asia and the Middle East. "But it is not enough, it argued. We would like to see China upgrade its commitments to current MTCR levels and implement effective export controls."

[179] Director of Central Intelligence, *Unclassified Report to Congress, January – June 1998*, Washington, DC

[180] For example, U.S. Congress, Senate Committee on Armed Services, *Worldwide Threats*, 106th Cong., 1st sess., February 2, 1999, S.Hrg. 106-436 (Washington: GPO, 2000); U.S. Congress, Senate Select Committee on Intelligence, *Current and Projected National Security Threats to the United States*, 106th Cong., 2nd sess., February 2, 2000, S.Hrg. 106-580 (Washington: GPO, 2000); Director of Central Intelligence, *Unclassified Report to Congress on the Acquisition of Technology Relating to Weapons of Mass Destruction and Advanced Conventional Munitions, 1 July through 31 December 1999*, Washington, DC; and Director of Central Intelligence, *Unclassified Report to Congress on the Acquisition of Technology Relating to Weapons of Mass Destruction and Advanced Conventional Munitions, 1 January through 30 June 2000*, Washington, DC.

[181] Deputy Director of National Intelligence for Analysis, *Unclassified Report to Congress on the Acquisition of Technology Relating to Weapons of Mass Destruction and Advanced Conventional Munitions, 1 January through 31 December 2010*, Washington, DC.

[182] Director of National Intelligence, James R. Clapper. *Unclassified Statement for the Record on the Worldwide Threat Assessment of the U.S. Intelligence Community for the Senate Select Committee on Intelligence*, January 31, 2012.

[183] This unreleased UN report is identified as: United Nations, *Final Report*, Panel of Experts Established Pursuant to resolution 1929 (2010), May 2010, pp. 28-29. The only online copy that could be found is: http://www.scribd.com/doc/55737041/Leaked-UN-Panel-of-Experts-Report-on-Iran-Sanctions-May-2011.

[184] Used for the production of motor exit cones and nozzles, reentry vehicle nose tips, heat shields and leading edges of control surfaces.

[185] Used to produce solid motor cases, interstages, wings, inlets, nozzles, heat shields, nose tips, structural members and frames.

[186] Used for producing solid-propellant rocket fuel.

[187] An oxidizing agent used in most modern solid-propellant formulas.

[188] An important fuel component of composite propellants in solid rocket motors.

[189] Director of Central Intelligence, *Unclassified Report to Congress on the Acquisition of Technology Relating to Weapons of Mass Destruction and Advanced Conventional Munitions, 1 January through 30 June 1999*, Washington, DC.

[190] *Congressional Record*, December 22, 2010, pp. S10984-S10985.

[191] Alon Ben-David, "West Doubts Iran Missile Claims," *Jane's Defence Weekly*, April 12, 2006.

[192] Douglas Barrie, "Firing Away; Missile in Iranian exercises underscores extent of guided-weapons ties between Tehran and Beijing," *Aviation Week and Space Technology*, April 10, 2006, p. 33.

[193] Bill Gertz, "Pentagon calls Iran missile claim and exaggeration," *The Washington Times*, April 4, 2006, p. A03.

[194] "Iran fires first long-range missiles in war games," *Agence France Presse—English*, November 2, 2006.

[195] "Iran ready to give missile systems to regional allies - army commander," *BBC monitoring Middle East - Political*, November 8, 2006, source: *Al-Alam TV*, Tehran (in Arabic), November 5, 2006.

[196] Reuters, "Iran Would Share Missiles," *The New York Times*, November 7, 2006, p. 6.

[197] Ibid.

[198] Iran claimed the Shahab-3 was fitted with a warhead containing 1,400 bomblets and that they landed at an intended target only a few meters away leaving a large crater larger than 20-30 meters. It would be used against large garrisons, large gatherings, aircraft carriers, etc. See "Iran's Guard Commander comments on Tehran's missile power," *BBC Monitoring Middle East - Political*, November 13, 2006, source: *Vision of the Islamic Republic of Iran Network 2*, Tehran, in Persian, November 12, 2006.

[199] Ed Blanche, "Iran stages display of missile firepower," *Jane's Missiles and Rockets*, January 1, 2007.

[200] "Report: Iran Test fires 3 New Missiles," *Associated Press Online*, November 3, 2006, on-line edition.

[201] David Shelby, "U.S. Admiral Condemns Iranian Missile Tests," *State Department Documents and Publications*, November 6, 2006. For Adm. Walsh, the concern expressed was not so much focused on this particular test, but the these missiles generally posed to the Gulf.

[202] David A. Fulghum, "Warhead Diplomacy," *Aviation Week & Space Technology*, July 14, 2008, p. 60.

[203] Pamela Hess, "Official: Iran missile tests used 'old equipment,'" *Associated Press*, July 12, 2008.

[204] See for instance, "Defiant Iran test-fires longer range missile," *Agence France Presse - English*, July 9, 2008.

[205] Pauline Jelinek, "Pentagon plumbing Iran's missile tests for clues," *The Associated Press*, July 9, 2008.

[206] Ali Akbar Dareini, "Iran tests long- and medium-range missiles in Strait of Hormuz," *Associated Press Worldstream*, July 9, 2008.

[207] "Iran fired only one missile on second day: U.S.," *Agence France Presse*, July 10, 2008.

[208] "Iran doctored missile test-firing photo: defence analyst," *Agence France Presse - English*, July 9, 2008, and Al Kamen, "Iran Apparently in Possession of Photoshop," *The Washington Post*, July 11, 2008, p. A15, Suburban Edition.

[209] Ibid.

[210] "The U.S. Department of State: Daily Press Briefing July 10, 2008," *States News Services*, July 10, 2008.

[211] Alan Crowell and Nazila Fathi, "Iran Test-Fires Missiles That Put Israel in Range," *The New York Times*, September 28, 2009, On-line edition.

[212] Ibid.

[213] William Branigin, Thomas Erdbrink, and Walter Pincus, "Iran Test Fires Its Most Advanced Missiles," *The Washington Post*, September 28, 2009, On-line edition.

[214] Jay Deshmukh and Hiedeh Farmani, "Iran launches missile war games," *Agence France Presse - English*, September 27, 2009.

[215] Iran Watch, Tracking Iran's Weapons of Mass Destruction Program, Wisconsin Project on Nuclear Arms Control, Iran Missile Milestones (for September 2009), http://www.iran watch.org/wmd/wponac-missilemilestones.htm

[216] Jay Deshmukh and Hiedeh Farmani, "Iran launches missile war games," *Agence France Presse - English*, September 27, 2009.

[217] "Iran Guards fire five missiles: TV," *Agence France Presse - English*, April 25, 2010, Newswire.

[218] "Iran test-fires powerful missiles in Persian Gulf drill: report," *Xinhua General News Service - English*, April 25, 2010, newswire.

[219] Various Iranian media and other reports provided differing designations.

[220] "Commander: All US, Israeli Bases within Iran's Missile Range," *FARS News Agency (Iran)*, June 28, 2011.

[221] Annual Report on Military Power of Iran, Department of Defense, April 2012, p. 4.

[222] "In 'Great Prophet VI' Maneuvers, Iran fires 14 ballistic missiles; IRGC Dep. Cmdr: Iran won't initiate attack – but response will be 'purely aggressive.'" MEMRI (Middle East Media Research Institute) Blog, June 28, 2011 http://www.thememriblog.org /blog_ personal/en/38864.htm

[223] William J. Broad, "Iran Unveils Missile Silos as it Begins War Games," *The New York Times*, June 27, 2011, Online edition.

[224] William J. Broad, "Iran shows off strengthened underground silos for missiles," *The International Herald Tribune*, June 29, 2011, p. 5.

[225] Ibid.

[226] "Iran TV shows footage of 'underground missile silo'—full text," *BBC Monitoring Middle East - Political*, June 28, 2011, source: Vision of the Islamic Republic Network 2, Tehran, in Persian, June 27, 2011.

[227] Lauren Gelfand, "Iran tests missiles amid suspicion of nuclear goals," *Jane's Defence Weekly*, June 30, 2011.

[228] See footnote 56.

[229] Thomas Erdbrink, "Iran test-fires missiles, shows secret silos," *The Washington Post*, June 28, 2011, on-line edition.

[230] The Persian Gulf missile is also known as the Khalije Fars Anti-Ship Ballistic Missile.

[231] "IRGC Starts Wargames by Firing Tens of Missiles," *FARS News Agency (Iran)*, July 3, 2012, English.

[232] Ibid. See also "Iranian Media Says Missile Able to Hit Israel Tested," *Radio Free Europe. Radio Liberty*, July 3, 2012, on-line edition.

[233] Associated Press, "Iranian reports: Military maneuvers display better accuracy, firing capability of missiles," *Washington Post*, July 13, 2012, On-line edition.

[234] Lt. Col (ret) Michael Segall (Israeli Defense Force), "Iran's Latest Missile Exercise: A Message of Force Projection," *Jerusalem Center for Public Affairs*, vol. 12, no. 16 (July 17, 2012), pp. on-line edition.

[235] This section was drawn from previous work by the author and Dianne Rennack, Specialist in Foreign Policy Legislation, CRS.

In: Iran's Ballistic Missile and Space Launch Programs ISBN: 978-1-62417-988-4
Editor: Steven Wietman

Chapter 2

IRAN: U.S. CONCERNS AND POLICY RESPONSES*

Kenneth Katzman

SUMMARY

A priority of Obama Administration policy has been to address the perceived threat posed to a broad range of U.S. interests by Iran, in particular by its advancing nuclear enrichment program. These advances have caused the government of Israel to assert that it might take unilateral military action against Iran's nuclear facilities unless the United States provides assurances that it will act, militarily if necessary, to prevent Iran from taking the final steps toward developing a nuclear weapon. Aside from the nuclear issue, the United States has long seen Iran's support for militant groups in the Middle East and in Iraq and Afghanistan as efforts to undermine U.S. interests and allies. U.S. officials also accuse Iran of helping Syria's leadership try to defeat a growing popular opposition movement and of taking advantage of Shiite majority unrest against the Sunni-led, pro-U.S. government of Bahrain.

The Obama Administration has orchestrated broad international pressure on Iran through economic sanctions, while also offering Iran sustained engagement if it verifiably demonstrates to the international community that its nuclear program is peaceful. Iran's leaders returned to nuclear talks with six powers in April 2012 after a one year hiatus. Three rounds of talks held in April, in May, and in June yielded no breakthroughs, but did explore a potential compromise under which Iran might end uranium enrichment to 20% purity (a level not technically far from weapons grade) in exchange for some sanctions relief. Subsequent technical talks produced no immediate progress, but might have set the stage for more high level talks in late 2012 or early 2013. The Administration asserts that there is time for accumulated international sanctions to compel Iran's leaders to compromise before U.S. military action be considered. Since the beginning of 2012, as significant multilateral sanctions have been added on Iran's oil exports— including an oil purchase embargo by the European Union that went into full effect on July 1, 2012—the regime has begun to acknowledge significant economic pressure. Those pressure led to a virtual collapse in the

* This is an edited, reformatted and augmented version of the Congressional Research Service Publication, CRS Report for Congress RL32048, dated November 15, 2012.

market value of Iran's currency, the rial, in early October 2012. Since late 2011, both Britain and Canada, have closed their embassies in Iran. Some reports suggest that the Administration might consider bilateral talks with Iran if doing so could head off confrontation.

The Administration and many outside experts also perceive that the legitimacy and popularity of Iran's regime is in decline, although not to the point where the regime's grip on power is threatened. There are few signs that the opposition in Iran or in exile have gained traction against the regime, even though international sanctions are causing clear public frustration. The reformist boycott of the March 2, 2012, parliamentary elections rendered the election a contest between factions supporting either President Mahmoud Ahmadinejad or Supreme Leader Ali Khamene'i. Khamene'i supporters were elected overwhelmingly. It is likely that only hardliners will be significant candidates in the next presidential election to be held on June 14, 2013.

The 112th Congress has supported additional economic sanctions against Iran, most recently with enactment of the Iran Threat Reduction and Syria Human Rights Act of 2012 (H.R. 1905, P.L. 112-158), which expands sanctions against companies that conduct energy and financial transactions with Iran. Additional sanctions were imposed by the European Union in mid-October and some Members of Congress reportedly are considering legislation that would compel a near- total global trade embargo on Iran.

POLITICAL HISTORY

Iran is a country of about 75 million people, located in the heart of the Persian Gulf region. The United States was an ally of the late Shah of Iran, Mohammad Reza Pahlavi ("the Shah"), who ruled from 1941 until his ouster in February 1979. The Shah assumed the throne when Britain and Russia forced his father, Reza Shah Pahlavi (Reza Shah), from power because of his perceived alignment with Germany in World War II. Reza Shah had assumed power in 1921 when, as an officer in Iran's only military force, the Cossack Brigade (reflecting Russian influence in Iran in the early 20th century), he launched a coup against the government of the Qajar Dynasty. Reza Shah was proclaimed Shah in 1925, founding the Pahlavi dynasty. The Qajars had been in decline for many years before Reza Shah's takeover. That dynasty's perceived manipulation by Britain and Russia had been one of the causes of the 1906 constitutionalist movement, which forced the Qajars to form Iran's first Majles (parliament) in August 1906 and promulgate a constitution in December 1906. Prior to the Qajars, what is now Iran was the center of several Persian empires and dynasties whose reach shrunk steadily over time. Since the 16th century, Iranian empires lost control of Bahrain (1521), Baghdad (1638), the Caucasus (1828), western Afghanistan (1857), Baluchistan (1872), and what is now Turkmenistan (1894). Iran adopted Shiite Islam under the Safavid Dynasty (1500-1722), which ended a series of Turkic and Mongol conquests.

The Shah was anti-Communist, and the United States viewed his government as a bulwark against the expansion of Soviet influence in the Persian Gulf and a counterweight to pro-Soviet Arab regimes and movements. Israel maintained a representative office in Iran during the Shah's time and the Shah supported a peaceful resolution of the Arab-Israeli dispute. In 1951, under pressure from nationalists in the Majles (parliament) who gained strength in the 1949 Majles elections, he appointed a popular nationalist parliamentarian, Dr. Mohammad Mossadeq, as prime minister. Mossadeq was widely considered left-leaning, and the United States was wary of his policies, which included his drive for nationalization of the oil industry, which had since 1913 been under the control of the Anglo-Persian Oil Company. Mossadeq's followers began an uprising in early August 1953 when the Shah tried to dismiss Mossadeq, and the Shah fled. The Shah was restored in a CIA-supported uprising that toppled Mossadeq ("Operation Ajax") on August 19, 1953.

The Shah tried to modernize Iran and orient it toward the West, but in so doing he alienated religious Iranians and the Shiite clergy and he allegedly tolerated severe repression and torture of dissidents by his SAVAK intelligence service. The Shah exiled Ayatollah Ruhollah Khomeini in 1964 because of Khomeini's active opposition, which was based on the Shah's anti-clerical policies and what Khomeini alleged was the Shah's forfeiture of Iran's sovereignty to the United States. Khomeini fled to and taught in Najaf, Iraq, a major Shiite theological center that contains the Shrine of Imam Ali, Shiism's foremost figure. There, he was a peer of senior Iraqi Shiite clerics and, with them, advocated direct clerical rule or *velayat-e-faqih* (rule by a supreme Islamic jurisprudent). In 1978, three years after the March 6, 1975, Algiers Accords between the Shah and Iraq's Baathist leaders, which settled territorial disputes and required each party to stop assisting each other's oppositionists, Iraq expelled Khomeini to France, from which he stoked the Islamic revolution. Mass demonstrations and guerrilla activity by pro-Khomeini forces, allied with a broad array of anti-Shah activists, caused the Shah's government to collapse in February 1979. Khomeini returned from France on February 1, 1979 and, on February 11, 1979, he declared an Islamic Republic of Iran. His political system of *velayat-e-faqih* was enshrined in the constitution that was adopted in a public referendum in December 1979 (and amended in 1989); it provided for the post of Supreme Leader. The regime based itself on strong opposition to foreign, particularly Western, influence, and relations between the United States and the Islamic Republic turned openly hostile after the November 4, 1979, seizure of the U.S. Embassy by pro-Khomeini radicals. Ayatollah Khomeini died on June 3, 1989, and was succeeded as Supreme Leader by Ayatollah Ali Khamene'i.

The regime faced serious unrest in its first few years, including a June 1981 bombing at the headquarters of the Islamic Republican Party (IRP) and the prime minister's office that killed several senior leaders, including Khomeini confidant Mohammad Hossein Beheshti. These events, along with the hostage crisis with the United States, provided cover for the regime to purge many of the secular, liberal, and left-wing personalities and parties in the anti-Shah coalition. Examples included the Tudeh Party (Communist), the People's Mojahedin Organization of Iran (PMOI, see below), the first elected President Abolhassan Bani Sadr, and the Iran Freedom Movement of the regime's first Prime Minister Mehdi Bazargan and, later, Ibrahim Yazdi. The regime was under economic and military threat during the 1980-1988 Iran-Iraq War, which resulted at times in nearly halting Iran's oil exports. Still, during these years, there was diversity of opinion in ruling circles.

Despite these struggles, during 1982 until 2009, the regime had faced only episodic, relatively low-level unrest from minorities, intellectuals, students, labor groups, and women. Since the June 2009 presidential election, the regime has struggled to contain popular dissatisfaction. In late 2009, several Iran experts believed this opposition movement—calling itself "The Green Path of Hope" or "Green Movement" (*Rah-e-Sabz*)—posed a serious challenge to the current regime. The regime subsequently pushed the Green Movement underground through imprisonment or house arrests of its leaders or main activists.

REGIME STRUCTURE, STABILITY, AND OPPOSITION

Iran's Islamic regime, established in a constitution adopted in a popular referendum, is widely considered authoritarian but not "one-man rule." The system provides for a degree of popular input and checks and balances provided by elected institutions. The Supreme Leader is not directly elected by the population; the president and the Majles (parliament) are. There are also direct elections for municipal councils, which in turn select mayors. Even within the unelected institutions, factional disputes between those who insist on ideological purity and those considered more pragmatic have been frequent and highly consequential.

Unelected Governing Institutions: The Supreme Leader, His Powers, and Other Ruling Councils

At the apex of the Islamic Republic's power structure is a "Supreme Leader" who has vast formal powers and no term limits. He is chosen by an elected body—the Assembly of Experts—which also has the constitutional power to remove him. Upon Ayatollah Khomeini's death, the Assembly selected one of his disciples, Ayatollah Ali Khamene'i, as Supreme Leader.[1] Although he has never had Khomeini's undisputed political or religious authority, the powers of the office have enabled Khamene'i to ensure that he is Iran's paramount leader. Formally, the Supreme Leader is commander-in-chief of the armed forces, giving him the power to appoint commanders and to be represented on the highest national security body, the *Supreme National Security Council*, composed of top military and civilian security officials. The Supreme Leader also has the power, under the constitution, to remove the elected president if either the judiciary or the elected *Majles* (parliament) decide that the president should be removed, with cause.

Still, the growing dependence of the regime on internal security forces caused Secretary of State Clinton to assert in February 2010 that the Supreme Leader's authority is being progressively usurped by regime security forces, most notably the Islamic Revolutionary Guard Corps (IRGC). This assessment is disputed by several outside experts who continue to see the Supreme Leader and his allies in the clerical establishment in firm control of regime decision making.

Council of Guardians and Expediency Council

The Supreme Leader appoints half of the 12-member *Council of Guardians*;[2] and the head of Iran's judiciary (currently Ayatollah Sadeq Larijani). Headed by Ayatollah Ahmad Jannati, the conservative-controlled Council of Guardians reviews legislation to ensure it conforms to Islamic law, and it screens election candidates and certifies election results. The Supreme Leader appoints members of the 42-member "Expediency Council," set up in 1988 to resolve legislative disagreements between the *Majles* and the Council of Guardians. The Expediency Council's powers were expanded in 2006 to include oversight of the executive branch (cabinet) performance. Its members serve five-year terms; its chairman, Ali Akbar Hashemi-Rafsanjani, was re-appointed in February 2007 and again in March 2012. The March 2012 reappointment defied predictions of many experts that he would be removed because of perceived disloyalty to the Supreme Leader, and was widely interpreted as a Khamene'i effort to keep Rafsanjani from supporting reformist leaders. Earlier, Rafsanjani was removed in March 2011 as head of the Assembly of Experts (see below). The Expediency Council's executive officer is former Revolutionary Guard commander-in-chief Mohsen Reza'i.

Elected Institutions: The Presidency, the Majles (Parliament), the Assembly of Experts, and Recent Elections

Elections in Iran have always lacked some credibility for international observers because hardliners are able to use their control over key election administration bodies such as the Interior Ministry and the Council of Guardians to limit the number and ideological diversity of candidates. The Council of Guardians has the power to approve or deny candidates based on its application of constitutional requirements about a candidate's knowledge of Islam and loyalty to the Islamic system of government.

Table 1. Supreme Leader: Grand Ayatollah Ali Khamene'i

Born in July 1939 to an Azeri (Turkic) family from Mashhad. Was jailed by the Shah of Iran for supporting Ayatollah Khomeini's revolution. After the regime took power in 1979, helped organize Revolutionary Guard and other security organs. Lost use of right arm in assassination attempt in June 1981. Was elected president during 1981-1989 and was selected Khomeini's successor in June 1989 upon his death. Upon that selection, Khamene'i religious ranking was advanced in official organs to "Grand Ayatollah" from the lower ranking "Hojjat ol-Islam." But, still lacks the undisputed authority and public adoration Khomeini had. Like Khomeini, Khamene'i has tended to intervene primarily to resolve factional disputes, or to quiet popular criticism, but has taken more of a day-to-day role since 2009 uprising. Has sided decisively with hardline opponents of Ahmadinejad since mid-2011.
Has taken consistently hard-line stances on foreign policy and particularly toward Israel, often calling it a cancerous tumor that needs to be excised from the region. Reputedly issued religious proclamation (2003) against Iran acquiring a nuclear weapon, and has publicly called doing so a "sin," but has consistently opposed bowing to any U.S. pressure on the nuclear issue. Yet, he is widely believed to fear direct military confrontation with United States on Iranian soil. Generally does not meet with Western officials and is suspicious of relations with the West as opening Iran to undue Western cultural influence, spying, and possible regime destabilization efforts. Opposes opening comprehensive direct talks with the United States, but has made public reference to letters to him from President Obama asking for renewed U.S.-Iran relations. On economic issues, he has tended to support the business community (bazaaris), and opposed state control of the economy, but believes Iran's economy is self-sufficient enough to withstand the effects of international sanctions.
His office is run by Mohammad Mohammadi Gopayegani, with significant input from Khamene'i's second son, Mojtaba, who is said to be acquiring increasing influence. Also advised by Keyhan editor Hossein Shariatmadari and former Foreign Minister Ali Akbar Velayati, who reportedly may run for President in June 2013 with Khamene'i's backing.
Potential successors include Ayatollah Ali Akbar Hashemi-Rafsanjani, Ayatollah Mohammad Taqi Mesbah Yazdi, Council of Guardians head Ayatollah Ahmad Jannati, Judiciary head Ayatollah Sadeq Larijani, and former Majles Speaker Gholam Haddad Adel. None is considered a clear consensus candidate should Khamene'i leave the scene unexpectedly.

Source: CRS.

Another criticism of the political process in Iran is the relative absence of political parties; establishing a party requires the permission of the Interior Ministry under Article 10 of Iran's constitution. The standards to obtain approval are high: to date, numerous parties have filed for permission since the regime was founded, but only those considered loyal to the regime have been granted (or allowed to retain) license to operate. Some of those authorized include Ahmadinejad's "Alliance of Builders of Islamic Iran" party and the "Executives of Construction" party. Some have been licensed and then banned, such as the two reformist parties, Islamic Iran Participation Front and the Organization of Mojahedin of the Islamic Revolution, which were formally outlawed in September 2010.

The Presidency

The main directly elected institution is the presidency, which is clearly subordinate to the Supreme Leader and most presidents during the Islamic republic have been unsuccessful in expanding their authority relative to the Supreme Leader. Still, the presidency is a coveted position which provides vast opportunities for the holder of the post to empower his political base and to affect day-to-day policy, particularly on economic issues. The president appoints and supervises the work of the cabinet, but the Supreme Leader is believed to have significant input into security-related cabinet appointments, including ministers of defense, interior, and intelligence (Ministry of Information and Security, MOIS). Prior to 1989, Iran had both an elected president as well as a prime minister selected by the elected Majles (parliament). However, the officials who held these posts during 1981-89 (Ali Khamene'i, who is now Supreme Leader, and Mir Hossein Musavi, who is now the main opposition leader, respectively) were in constant institutional conflict and the constitution was revised in 1989 to eliminate the post of prime minister.

As the top governing official, the presidency is also responsible for oversight bodies, including the Anticorruption Headquarters and the General Inspection Organization. The presidency develops the budgets of cabinet departments and imposes and collects taxes on corporations and other bodies. However, implementation of all these functions is said to be uneven, and presidential authority is often undermined by key clerics and allies of the Islamic Revolutionary Guard Corps (IRGC) and other powerful institutions. All government officials are required to submit annual financial statements to state auditors, but there is no confirmation that such procedures are followed. Religious foundations, called "bonyads," for example are tax exempt and their import-export operations often unregulated. Through profits earned from its affiliate companies, the IRGC is widely known to spend funds, outside its approved government budget, on arms, technology, support to pro-Iranian movements, and other functions. Such expenditures are conducted with almost no official oversight.

In a speech on October 16, 2011, Supreme Leader Khamene'i raised the possibility of eliminating the post of president and restoring the post of prime minister. The comments were viewed in the context of a rift between him and President Ahmadinejad, discussed below. Khamene'i indicated the change would not be difficult to orchestrate—although it would require a change to the constitution. In late July 2012, a parliamentary committee was established to assess the possibility of such a change, although observers say it will not be implemented before the June 2013 election and the election will go forward.

The Majles

Iran's Majles, or parliament, consists of 290 seats, all elected. However, there are reserved seats (one each) for members of Iran's religious minorities, including Jews and Christians. There is no "quota" for the number of women to be elected, but women regularly run and win election, although not in proportion to their percentage of the population. Majles elections occur one year prior to the presidential elections; the elections for the ninth Majles were held on March 2, 2012, and the dynamics and outcome of the upcoming contest are discussed below.

Cabinet appointments are subject to confirmation by the Majles (parliament), which also drafts and acts on legislation. The unicameral Majles in Iran is highly factionalized but, as an institution, it is far from the "rubber stamp" that characterizes many elected national assemblies in the region, but it generally has lost institutional disputes to the president. Among its main duties is to consider and enact a proposed national budget; that review typically takes place each February and March in advance of the Persian New Year (Nowruz) on March 21.

The Assembly of Experts

Another elected institution, mentioned above, is the *Assembly of Experts*. It is akin to an electoral college: it is empowered to choose a new Supreme Leader upon the death of the incumbent, and it oversees the work of the Supreme Leader. The Assembly can replace him if necessary, although invoking that impeachment power would, in most circumstances, be highly controversial. It is also the body empowered to amend the constitution. The Assembly has 86 seats, elected to an eight-year term, with elections conducted on a provincial basis. It generally meets two times a year, for a few days each. The fourth election for the Assembly was held on December 15, 2006; after that election, Rafsanjani, still a major figure having served two terms as president (1989-1997), was named deputy leader of the Assembly. After the death of the leader of the Assembly (Ayatollah Meshkini), Rafsanjani was selected its head in September 2007. However, as part of the broader power struggles within the regime that have raged since the post-2009 election uprising, Rafsanjani was not reelected as Assembly of Experts chair in March 2011. He was replaced by aging and infirm compromise candidate Ayatollah Mohammad Reza Mahdavi-Kani. See **Figure 1** for a chart of the Iranian regime.

Recent Elections: First Ahmadinejad Election in 2005

After suffering several presidential election defeats at the hands of President Mohammad Khatemi and the reformists in the 1997 and 2001 presidential elections, hardliners successfully moved to regain the sway they held when Khomeini was alive. Conservatives won a majority (155 out of the 290 Majles seats) in the February 20, 2004, Majles elections (which are always held one year prior to each presidential election), in large part because of the Council of Guardians' disqualification of 3,600 reformist candidates, including 87 Majles incumbents. The George W. Bush Administration and the Senate (S.Res. 304, adopted by unanimous consent on February 12, 2004) criticized the elections as unfair because of the disqualifications.

As the reformist faction suffered setbacks, the Council of Guardians narrowed the field of candidates for the June 2005 presidential elections to 8 out of the 1,014 persons who filed.

Rafsanjani[3] was considered the favorite against several opponents more hardline than he is—three had ties to the Revolutionary Guard: Ali Larijani; Mohammad Baqer Qalibaf; and Tehran mayor Mahmoud Ahmadinejad. In the June 17, 2005, first round, turnout was about 63% (29.4 million votes out of 46.7 million eligible voters). With 21% and 19.5%, respectively, Rafsanjani and Ahmadinejad, who did unexpectedly well because of tacit backing from Khamene'i, moved to a runoff. Reformist candidates (Mehdi Karrubi and Mostafa Moin) fared worse than expected. Ahmadinejad won in the June 24 runoff, receiving 61.8% to Rafsanjani's 35.7%. He took office on August 6, 2005.

During his first term, splits widened between Ahmadinejad and other conservative members of his "Principalist" (*usulgaran*) faction. That rift was evident in the March 2008 Majles elections in which his base of support fractured and some conservatives ran as an anti-Ahmadinejad bloc. These splits foreshadowed the broader rift with the Supreme Leaders, discussed below.

Disputed Ahmadinejad 2009 Reelection: Protests Erupt and Second Term is Riven by Schisms

With splits in Ahmadinejad's base, prospects for reformists to unseat Ahmadinejad through the established election process seemed to brighten. In February 2009, when Khatemi indicated a willingness to run, but he ultimately yielded to and endorsed a fellow reformist, Mir Hossein Musavi. Musavi was viewed as somewhat less divisive (and therefore more acceptable to the Supreme Leader) than Khatemi because of Musavi's service as prime minister during the 1980-1988 Iran-Iraq War.

A total of about 500 candidates for the June 12, 2009, presidential elections registered their names during May 5-10, 2009. The Council of Guardians decided on four final candidates on May 20: Ahmadinejad, Musavi, Mehdi Karrubi, and former Commander-in-Chief of the Revolutionary Guard Mohsen Reza'i. The Interior Ministry, which runs the election, also instituted an unprecedented series of one-on-one debates, which including Ahmadinejad's acrimonious accusations of corruption against Rafsanjani and against Musavi's wife. If no candidate received more than 50% of the vote on June 12, there would have been a runoff one week later.

The challengers and their backgrounds and platforms were as follows.

- Mir Hosein Musavi. The main reformist candidate. See **Table 2**.
- Mehdi Karrubi. See **Table 2**.
- Mohsen Reza'i. As noted, commander in chief of the Revolutionary Guard through the Iran-Iraq War. About 58 years old, he was considered an anti- Ahmadinejad conservative. Reza'i dropped out just prior to the 2005 presidential election. He alleged fraud in the 2009 election but later dropped his challenge.

The outcome of the election was always difficult to foresee; polling was inconsistent. Musavi supporters using social media such as Facebook and Twitter organized large rallies in Tehran, but pro-Ahmadinejad rallies were large as well. Turnout was high at about 85%; 39.1 million valid (and invalid) votes were cast. The Interior Ministry announced two hours after the polls closed that Ahmadinejad had won, although in the past results have been announced the day after. The totals were announced on Saturday, June 13, 2009, as follows.

- Ahmadinejad: 24.5 million votes—62.6%
- Musavi: 13.2 million votes—33.75%
- Reza'i: 678,000 votes—1.73%
- Invalid: 409,000 votes—1%
- Karrubi: 333,600 votes—0.85%

Almost immediately after the results of the election were announced on June 13, 2009, Musavi supporters began protesting the results, citing: the infeasibility of counting 40 million votes so quickly and the barring of candidate observers at many polling stations. Khamene'i declared the results a "divine assessment," appearing to certify the results even though formal procedures require a three-day complaint period. Some outside analysts said the results tracked pre-election polls, which showed strong support for Ahmadinejad in rural areas and among the urban poor.[4]

"Green Movement" Protest Movement Forms

Continuing to use Facebook and Twitter, and fueled by outrage over regime use of force as depicted on YouTube, the demonstrations built throughout June 13-19, 2009, largely in Tehran but also in other cities. Security forces used varying amounts of force to control them, causing 27 protester deaths (official tally) during that period, with opposition groups reporting over 100 killed, including a 19 year-old woman Neda Soltani, who subsequently became an emblem of the opposition movement. The protesters' hopes of having Khamene'i annul the election were dashed by his major Friday prayer sermon on June 19, 2005 in which he refuted allegations of vast fraud and threatened a crackdown on further protests. Protesters continued despite Khamene'I's warning. On June 29, 2009, the Council of Guardians tried to address the complaints by performing a televised recount of 10% of the votes of Tehran's districts and some provincial ballots and, finding no irregularities, certified the results. As 2009 progressed, the opposition congealed into the "Green Movement of Hope and Change," which later moved beyond the election issue into a challenge to the regime, as discussed below.

Ahmadinejad's Second Term: Divisions within the Regime Increase

As the Green Movement gathered strength in 2009, splits within the regime widened, although most of the core regime leaders tried to remain outwardly unified. As unrest faded from the streets at the start of 2010, Ahmadinejad sought to promote the interests of his loyalists and promote what his critics say is a nationalist version of Islam that limits the authority of Iran's clerics. Ahmadinejad's perceived favorite has been his former chief-of-staff, Esfandiar Rahim Mashai, to whom he is related through their children's marriage This caused anti-Ahmadinejad hardliners to rally around the Supreme Leader Khamene'i—who himself is believed suspicious of Ahmadinejad's allies' ambitions and ideology—to try to weaken Ahmadinejad. Many in the regime want to see antagonists of his, particularly moderate-conservatives such as Ali Larijani or Mohammad Baqr Qalibaf, as the next president in 2013.

Table 2. Major Factions, Personalities, and Interest Groups

Conservatives	
Supreme Leader Ali Khamene'i	See box above.
Expediency Council Chair Ayatollah Ali Akbar Hashemi-Rafsanjani	A longtime key regime strategist, Khomeini disciple, and advocate of "grand bargain" to resolve all outstanding issues with United States. Was Majles speaker during 1981-1989 and president 1989-1997. One of Iran's richest men, family owns large share of Iran's total pistachio production. Was seriously weakened in March 2011 by ouster as Assembly of Experts chairman, an outcome attributed to his tacit support of Green challenge to Ahmadinejad reelection. Rafsanjani funded much of Musavi's election campaign and criticized crackdown on Green protests. But, Khamene'i has rehabilitated him by reappointing him Expediency Council chair in March 2012. And, Rafsanjani appeared next to Khamene'I during September 2012 Non-Aligned Movement summit in Tehran. Political activities of his children remain a liability for Rafsanjani in regime circles. Daughter Faizah participated in several 2009 protests, was detained briefly in February 2011 for protesting, and was jailed in September 2012 for opposition activities. Five Rafsanjani family members arrested in June 2009 (and another briefly detained in March 2010), and there was a May 2010 threat to arrest his son, Mehdi, if he returns from exile in Britain.
President Mahmoud Ahmadinejad	See box below.
Majles Speaker Ali Larijani	Majles Speaker since 2008. Likely to run again for president in 2013. Former state broadcasting head (1994-2004) and minister of culture and Islamic guidance (1993), was head of Supreme National Security Council and chief nuclear negotiator from August 2005 until October 2007 resignation, and sought to avoid U.N. Security Council isolation. Is politically close to Khamene'i and a leading antagonist of Ahmadinejad. Brother of judiciary head. Another brother, Mohammad Javad, was deputy foreign minister (1980s) and now heads a government human rights body.
Tehran Mayor Mohammad Baqer Qalibaf	Former Revolutionary Guard Air Force commander and police chief, but a moderate-conservative ally of Larijani and critic of Ahmadinejad. Encourages comparisons of himself to Reza Shah, invoking an era of stability and strong leadership. Lost in 2005 presidential elections, but supporters won 9 out of 15 seats on Tehran city council in December 2006 elections, propelling him to become mayor of Tehran. Has won support for Tehran's cleanliness and relative order. Recruited moderate-conservatives for March 2008 Majles election. May run again in 2013.
Senior Shiite Clerics	The most senior clerics, most of whom are in Qom, including several Grand Ayatollahs, are generally "quietist"—they believe that the senior clergy should refrain from direct involvement in politics. These include Grand Ayatollah Nasser Makarem Shirazi, Grand Ayatollah (former judiciary chief) Abdol Karim Musavi- Ardabili, and Grand Ayatollah Yusuf Sanei, all of whom have criticized regime crackdown against oppositionists. Others believe in political involvement, including Ayatollah Mohammad Taqi Mesbah Yazdi. He is founder of the hardline Haqqani school, and has been considered spiritual mentor of Ahmadinejad, although he headed a bloc of hardliners not supportive of the president in the March 2, 2012, *Majles* elections. Yazdi, an assertive defender of the powers of the Supreme Leader and a proponent of an "Islamic state" rather than the current "Islamic republic," fared poorly in December 2006 elections for Assembly of Experts. Another hardline cleric is Ayatollah Kazem Haeri, mentor of Iraqi cleric Moqtada Al Sadr.
Judiciary Chief/Ayatollah Sadeq Larijani	Judiciary head since August 2009. Brother of Ali Larijani; both are close to the Supreme Leader and opponents of Ahmadinejad. Both support hard line against dissidents.
Militant Clergy Association	Longtime organization of hardline clerics headed by Ayatollah Mohammad Mahdavi-Kani, who became chair of the Assembly of Experts on March 9, 2011. Did not back Ahmadinejad for reelection in 2009 vote and led a bloc opposing Ahmadinejad in the March 2, 2012, *Majles* elections.

Conservatives	
Bazaar Merchants ("Bazaaris")	The urban bazaar merchants fear jeopardizing the economy by participating in political opposition activity; have conducted only a few strikes or other organized action since the 1979 revolution. The bazaaris are not a monolithic group; each city's bazaars are organized by industry (e.g., carpets, gold, jewelry, clothing) and bazaari positions tend to be reached by consensus among elders representing each industry represented at the bazaar.
Opposition/"Green Movement" (Rah-e-Sabz) All of the blocs and personalities below can be considered, to varying degrees, as part of the Green Movement. However, overall leadership of the movement and decision-making on protest activities is unclear, with several components competing for preeminence. Some Green supporters have left for Europe, Asia, or the United States.	
Titular Green Movement Leaders: Mir Hossein Musavi/ Mohammad Khatemi/Mehdi Karrubi	Khatemi—reformist president during 1997-2005 and declared he would run again for president in June 2009 elections, but withdrew when allied reformist Mir Hossein Musavi entered the race. Khatemi was elected president in May 1997, with 69% of the vote; reelected June 2001 with 77%. Rode wave of sentiment for easing social and political restrictions, but these groups became disillusioned with Khatemi's failure, as president, to stand up to hardliners on reform issues. Has hewed to staunch anti-Israel line of most Iranian officials, but perceived as open to accepting a Palestinian-Israeli compromise. Perceived as open to a political compromise that stops short of replacement of the regime, and voted in March 2, 2012, election, ignoring reformist boycott of that election. Now heads International Center for Dialogue Among Civilizations. Visited United States in September 2006 to speak at Harvard and the Washington National Cathedral on "dialogue of civilizations." Titular leader of the Green movement, Musavi is a non-cleric. About 68. An architect by training, and a disciple of Ayatollah Khomeini, he served as foreign minister (1980), then prime minister (1981-1989), at which time he successfully managed the state rationing program during the privations of the Iran-Iraq War but often feuded with Khamene'i, who was then president. At that time, he was an advocate of state control of the economy. His post was abolished in the 1989 revision of the constitution. Musavi later adopted views similar to Khatemi on political and social freedoms and on reducing Iran's international isolation, but supports strong state intervention in the economy to benefit workers, lower classes. Appeared at some 2009 protests, sometimes intercepted or constrained by regime security agents. However, not necessarily respected by harder line opposition leaders who criticize his statements indicating reconciliation with the regime is possible. He and wife (prominent activist Zahra Rahnevard) repeatedly harassed by regime during 2009 protests. He and Mehdi Karrubi, below, placed under strict house arrest after Green demonstrations resumed on February 14, 2011. With Karrubi, supported reformist boycott of March 2, 2012, Majles elections. A founder of the leftwing Association of Combatant Clerics (different organization but with similar name from that above), Mehdi Karrubi was Speaker of the Majles during, 1989-1992 and 2000-2004. Formed a separate pro-reform "National Trust" faction after losing 2005 election. Ran again in 2009, but received few votes and subsequently emerged, along with Musavi, as a leader of the Green Movement. Was physically blocked by regime from attending Green demonstrations during 2010 and, with Musavi, was put under house arrest as of February 14, 2011. Taken away to complete isolation (except for regime agents) at a two room office on July 16, 2011. Reportedly was allowed some access to his family in December 2011.
Student Opposition Leaders/ Office of Consolidation of Unity (Daftar Tahkim-e-Vahdat) and Other Groups	Groups composed of well-educated, Westernized urban youth are the backbone of the Green Movement. They are attempting, with mixed success, to gain support of older generation, labor, clerics, village-dwellers, and other segments. Many in the Office of Consolidation of Unity, the student group that led the 1999 riots but which has since become controlled by regime loyalists, believes that major reform of the current regime might be acceptable. An offshoot of the Office—the Confederation of Iranian Students (CIS)—believes in regime replacement and consists of pro-American, pro-free market activists. CIS co- founder, Arzhang Davoodi, convicted in August 2012 of "war against God" which carries life sentence. The other co-founder, Amir Abbas Fakhravar, is based in Washington, D.C. Another dissident figure who emerged from the 1999 student- riots is Ahmad Batebi.

Table 2. (Continued)

Conservatives	
Islamic Iran Participation Front (IIPF)	The most prominent and best organized pro-reform grouping, but has lost political ground to Green Movement groups advocating outright overthrow of the regime. Its leaders include Khatemi's brother, Mohammad Reza Khatemi (a deputy speaker in the 2000-2004 Majles) and Mohsen Mirdamadi. Backed Musavi in June 2009 election; several IIPF leaders, including Mirdamadi, detained and prosecuted in postelection dispute. The party was outlawed by the regime in September 2010.
Mojahedin of the Islamic Revolution Organization (MIR)	Composed mainly of left-leaning Iranian figures who support state control of the economy, but want greater political pluralism and relaxation of rules on social behavior. A major constituency of the reformist camp. Its leader is former Heavy Industries Minister Behzad Nabavi, who supported Musavi in 2009 election and has been incarcerated for most of the time since June 2009. The organization was outlawed by the regime simultaneously with the outlawing of the IIPF, above.
Combatant Clerics Association	Very similar name to organization above, but politically very different. Formed in 1988, it is run by reformist, not hardline, clerics and officials. Leading figures include Mohammad Khatemi, former Interior Minister Ali Akbar Mohtashemi-Pur, and former Prosecutor General Ali Asgar Musavi-Koiniha.
Labor Unions	Organized labor has suffered from official repression for many years. Organized labor is not a core constituency of the Green Movement, but laborers viewed as increasingly sympathetic to political change. Some labor protests took place in Tehran on "May Day" 2010, and selected small strikes (truckers, some factories) during 2010 led some experts to believe that labor might be gravitating toward Green Movement. However, younger Green Movement activists are suspicious of labor as a leftwing bastion. Others say union members fear income disruption if they openly defy the regime. A bus drivers' union leader, Mansur Osanloo, has been in jail since July 2007.
Other Prominent Dissidents	Other leading dissidents, some in Iran, others in exile, have been challenging the regime since well before the Green Movement formed and are now significant opposition figures. Journalist Akbar Ganji conducted hunger strikes to protest regime oppression; he was released on schedule on March 18, 2006, after sentencing in 2001 to six years in prison for alleging high-level involvement in 1999 murders of Iranian dissident intellectuals. Abdol Karim Soroush, now exiled, has challenged the doctrine of clerical rule. Former Revolutionary Guard organizer Mohsen Sazegara is based in the United States, but his role in the IRGC likely discredits him in the eyes of dissidents who want regime replacement. Other significant dissidents include former Culture Minister Ataollah Mohajerani, Mohsen Kadivar, and Fatemah Haghighatgoo. Some well known dissidents who remained in Iran and were arrested in 2010 include filmmaker Jafar Panahi and journalist Abdolreza Tajik. In November 2008, before the 2009 unrest, famed Iranian blogger Hossein Derakshan was jailed; he is serving a 20-year prison sentence. 80-year-old Iran Freedom Movement leader Ibrahim Yazdi was released from prison in April 2011 after resigning as the Freedom Movement's leader. One major dissident figure is Nobel Peace Prize laureate (2003) and Iran human rights activist lawyer Shirin Abadi. She has often represented clients persecuted or prosecuted by the regime but she left Iran for Europe, fearing arrest in connection with the 2009 postelection dispute. In December 2009, the regime confiscated her Nobel Prize. In April 2012, she publicly opposed U.S. and allied sanctions against Iran. In January 2011, a colleague, Nasrin Sotoudeh, was sentenced to 11 years in prison.

The infighting evolved into a rift between Ahmadinejad and Khamene'i, breaking out into the open in April 2011 when Ahmadinejad dismissed the intelligence minister Heydar Moslehi and attempted to replace him with a Mashai loyalist. The Supreme Leader reinstated Moslehi, and Ahmadinejad protested by refusing to attend cabinet meetings from April 24 to May 4, 2011. The political establishment rallied around the Supreme Leader, and Ahmadinejad accepted Moslehi's reinstatement.

Later in 2011, 25 Mashai loyalists were charged with witchcraft or sorcery. In September 2011, the split continued with allegations that a $2.6 billion embezzlement scheme involving fraudulent letters of credit were facilitated by Mashai—an implied link of the scam to Ahmadinejad himself. On February 7, 2012, the rift escalated further when the Majles, still mostly populated by those loyal to the Supreme Leader, voted to summon Ahmadinejad for formal questioning—the first time this has happened since the Islamic revolution. He made the appearance on March 14, 2012, after the March 2 Majles elections, but the session reportedly was not contentious. On July 30, 2012, four people were sentenced to death in the alleged embezzlement scheme, the first sentences of a total of 39 persons convicted in the case—the death sentences were considered a further rebuke to Ahmadinejad. On September 27, 2012, Ahmadinejad's press adviser, Ali Akbar Javanfekr, was sentenced to six months in jail for publishing materials critical of the Supreme Leader. That sentencing provoked another row in late October 2012 when Judiciary chief Sadeq Larijani (brother and ally of the Majles Speaker Ali Larijani) blocked Ahmadinejad from visiting the aide in prison, and Ahmadinejad asserted he did not need judiciary permission to do so. The eruption of this infighting in public led the Supreme Leader to threaten, on October 31, 2012, that public disputes among high officials could be treated as treason. A few days after that warning, the Majles voted to summon Ahmadinejad for questioning about the alleged mismanagement of the economy and the collapse of Iran's currency in early October, an issue discussed later. Ahmadinejad has until early December 2012 to respond to the Majles summons.

March 2, 2012, Majles Elections Marginalize Ahmadinejad

The 2012 Majles elections were held in the context of the schisms discussed above, and contributed to Ahmadinejad's current political weakness as his presidency winds down. Reflecting reduced faith in the fairness of the elections, during the candidate registration period, December 24-December 30, 2011, 5,400 Iranians put their names forward to compete - 33% fewer than those who filed candidacies four years ago. Only 10% were women. The leading reformist factions boycotted the elections, perceiving that the Council of Guardians was likely to limit voter choice to only hardline candidates. Perhaps justifying those fears, the Interior Ministry, the first body to screen candidates, disqualified 17% of the candidates as of January 24. The Council of Guardians, the ultimate arbiter, reinstated some of those candidates and issued the final candidate list of 3,400 (for the 290 seats) on February 21, 2012.

After the final candidate list was established, the regime used exhortations of nationalist obligations to try to encourage a large turnout. The reformist boycott left pro and anti-Ahmadinejad hardline factions to compete against each other. Ahmadinejad and his allies reportedly concentrated their efforts on rural areas where Ahmadinejad is relatively popular. The two pro-Khamene'i blocs—one centered around Ayatollah Mesbah-Yazdi (the Front of Stability of the Islamic Revolution) and another centered around Assembly of Experts chair Mahdavi-Kani (United Front of Conservatives, and included likely 2013 presidential candidates Qalibaf and Ali Larijani)—concentrated their efforts mainly in urban and suburban areas.

The regime announced a turnout of about 65%, which it asserted was a retort to international pressure and a sign of regime popularity. Khamene'i loyalists won a clear majority in it—and in a May 4, 2012, runoff for seats not determined—and control about 75% of the seats in the ninth *Majles*. Gholam Haddad Adel was nominated to return to the speaker role; his prominence is in part because of his relationship to Khamene'i. His daughter is married to Khamene'i's son, Mojtaba, who is one of his top aides. However, reflecting its confidence in Larijani, the Majles voted to retain Larijani as speaker in late May 2012. Following the Majles elections, many experts concluded that the Supreme Leader had consolidated his authority and rendered Ahmadinejad virtually irrelevant in his final year in office.

Mahmoud Ahmadinejad

First non-cleric to be president of the Islamic republic since the assassination of then-president Mohammad Ali Rajai in August 1981. About 58, he asserts he is a "man of the people," the son of a blacksmith who lives in modest circumstances, who would promote the interests of the poor and return government to the original principles of the Islamic revolution. Has burnished that image as president through regular visits to poor areas and through subsidies directed at the lower classes. His official biography says he served with the "special forces" of the Revolutionary Guard, and he served subsequently (late 1980s) as a deputy provincial governor. Although he is a member of the Builders of Islamic Iran party, he more closely identifies with a "Principalist" faction composed of former Guard and Basij (volunteer popular forces) leaders and other hardliners. U.S. intelligence reportedly determined he was not one of the holders of the 52 American hostages during November 1979-January 1981.

Other accounts say Ahmadinejad believes his mission is to prepare for the return of the 12th Imam—Imam Mahdi—whose return from occultation would, according to Twelver Shiite doctrine, be accompanied by the establishment of Islam as the global religion. Earned clerical criticism in May 2008 for again invoking intervention by Imam Mahdi in present day state affairs.

Following limited recount, declared winner of June 12, 2009, election. Well earlier, had been a controversial figure for inflammatory statements. He attracted significant world criticism for an October 26, 2005, Tehran conference entitled "A World Without Zionism" by stating that "Israel should be wiped off the map." In an October 2006 address, Ahmadinejad said, "I have a connection with God." He insisted on holding a December 2006 conference in Tehran questioning the Holocaust, a theme he has returned to several times since, including at a September 2007 speech at Columbia University. A U.N. Security Council statement and Senate and House resolutions (H.Res. 523 and S.Res. 292), passed by their respective chambers, condemned the statement. On June 21, 2007, the House passed H.Con.Res. 21, calling on the U.N. Security Council to charge Ahmadinejad with violating the 1948 Convention on the Prevention and Punishment of the Crime of Genocide; the Convention includes "direct and public incitement" of genocide as a punishable offense. On March 6, 2010, Ahmadinejad called the September 11, 2001, attacks on the United States a "big lie" used to justify intervention in Afghanistan. During his September 24, 2012, speech to the U.N. General Assembly in New York, he repeated past assertions that Israel must be eliminated. In interviews connected to that visit, he continued to criticize homosexuality and those who oppose any further research on or questioning of the Holocaust. Was apparent target of an unsuccessful grenade attack on his motorcade in the city of Hamedan on August 4, 2010. As noted, has steadily lost influence as a result of a power struggle with the Supreme Leader since early 2011.

Next Presidential Election

The date for the next presidential election has been set for June 14, 2012. The widely mentioned and favored candidates have been Ali Larijani and Mohammad Baqr Qalibaf, both well-known Khamene'i supporters. In August 2012, the Supreme Leader's foreign policy advisor, former Foreign Minister Ali Akbar Velayati, expressed interest in running. He is somewhat more moderate on most policies than either Larijani or Qalibaf, and has a long and close association with Khamene'i. It is likely that Ahmadinejad's ally Masha'i, discussed above, will not run—or be allowed to run—in the election. There has been some speculation that Rafsanjani, who has been rehabilitated politically by Khamene'i since mid-2012, might run again. Another potential candidate mentioned is Foreign Minister Ali Akbar Salehi.

A question under debate is whether any reformist figure might run. Some experts believe that the regime seeks additional legitimacy and will try to induce a reformist, such as former Interior Minister Abdollah Nuri, to run. A reformist is unlikely to win the election because many reformists are certain to boycott the vote. Others believe that the Supreme Leader and his supporters want only pliable allies to run and do not want to risk another uprising if a reformist leader runs and is declared the loser.

The Opposition

The popular uprising of 2009 constituted the most significant unrest faced by the regime since its inception in 1979. Many experts on Iran believe that the still seething opposition remains a key concern of the regime, particularly in the context of successful uprisings in the Arab world in 2011-2012 and the collapse of Iran's currency, the rial, in October 2012. Still, the regime's willingness to use force against unrest and the lack of clear leadership of the protest movement clouds the opposition's prospects to mount a sustained return to the streets. Not all the opposition operates under the Green Movement banner; some opposition groups in exile or in Iran operate separately, and others act to further ethnic or other interests.

The Green Movement and Its Uprising

The Green Movement, the genesis of which was the post-presidential election protests as discussed above, is centered around educated, urban youth, intellectuals, and former regime officials. Perhaps accounting for its failure to challenge the regime over the past two years, it has been unable to incorporated many traditionally conservative groups such as older Iranians and Iranians who live in rural areas. It remains divided between those who believe the regime can be reformed and moderated, and those who believe it must be replaced outright by a more secular system of government.

The year 2009 was clearly "the high water mark" of the Green Movement to date. After the initial post-election daily protests, Green Movement members organized protests around major holidays and called openly for the downfall of the regime, rather than its reform. Some of the protests in late 2009 nearly overwhelmed regime security forces. Large protests were held on the July 9 anniversary of the suppression of the 1999 student riots; the August 5, 2009, inauguration of Ahmadinejad; September 18, 2009 ("Jerusalem Day"); November 4, 2009, (30th anniversary of the takeover of the U.S. embassy in Tehran); and the Ashura holy day (December 27, 2009). The latter protest was marked by the seizure and burning of some police vehicles, and the refusal by some police to beat protesters; it spread to smaller cities and some clerics participated.

The movement's outward activity declined after its demonstration planned for the February 11, 2010, anniversary of the founding of the Islamic Republic (in 1979) was suppressed. With weeks to prepare, the regime limited opposition communication, made several hundred preemptive arrests, and executed some oppositionists in January 2010. Minor protests were held on March 16, 2010, a Zoroastrian holiday (Fire Festival), and there were scattered protests in major cities on May 1, 2010 (May Day). Musavi and Karrubi called for a huge demonstration on the June 12, 2010, anniversary of the election, but, sensing regime preparations for repression, the two publicly "called off" the protest in order to avoid harm to protesters.

A major question was whether the opposition uprisings in Tunisia and Egypt, which toppled leaders there in January and February 2011, would reinvigorate the Green Movement, which has used similar social media techniques and has similar grievances. The regime, seeking to parry such parallels, praised the Tunisian and Egyptian events as inspired by Iran's 1979 revolution, while Green Movement leaders compared those uprisings to their uprising in 2009. Musavi and Karrubi called for protests on February 14, 2011, and there were numerous clashes with tear-gas- wielding riot police in Tehran and other cities but, in advance of that demonstration, Karrubi and Musavi were placed under house arrest. Further weekly protests, which reportedly drew large numbers of protesters, were held from February 20, 2011 until Nowruz (March 21, 2011). But, no major demonstrations materialized at the 2011 second anniversary of the disputed election.

Despite these setbacks, observers in Iran say the Green Movement remains active underground. It conducted some significant protests on the February 14, 2012, anniversary of the February 14, 2011, protests. This occurred despite the January 2012 regime arrests of numerous journalists and bloggers. However, no additional protests erupted in the run-up to the March 2, 2012, Majles elections, in part because the Green Movement leaders boycotted the vote and their supporters did not have candidates to champion. The collapse of the rial in early October 2012 sparked major street demonstrations on October 3, 2012, the first demonstrations since early 2012, although the protests were linked to economic issues and did not necessarily reflect the more political agenda of the Green movement.

Exiled Opposition Groups: Supporters of the Son of the Late Shah of Iran

Some Iranian outside Iran, including in the United States, want to replace the regime with a constitutional monarchy led by Reza Pahlavi, the U.S.-based son of the late former Shah and a U.S.-trained combat pilot. The Shah's son, who is about 57 years old, has delivered statements condemning the regime for the post-2009 election crackdown and he has called for international governments to withdraw their representation from Tehran. He is supported by Iranian exile-run stations in California.[5] A younger brother, Ali Reza Pahlavi, committed suicide in January 2011.

Pahlavi has always had some support particularly in the older generation in Iran, but he reportedly is trying to broaden his following as perceptions of regime vulnerability have increased since 2009. Since March 2011, he has been increasingly cooperating with—and possibly attempting to co-opt—younger Green Movement figures. In a meeting with the author in June 2011, Pahlavi indicated that an internationally provided "strike fund" would help Iranian labor rise up against the regime by protecting their incomes from regime retaliation. He also advocated establishing a large scale opposition radio station, funded presumably by wealthy Persian Gulf states.. As of late 2012, he is said to be trying to assemble a "National Iranian Council" modeled on similar bodies representing the revolutions in Libya and Syria since early 2011.

Exiled Opposition Groups: People's Mojahedin

Some groups have been committed to the replacement of the regime virtually since its inception, and have used violence to achieve their objectives. Their current linkages to the Green Movement are tenuous, if existing at all, and some indications suggest these movements want to dominate any coalition that might topple the regime.

One of the best-known exiled opposition groups is the People's Mojahedin Organization of Iran (PMOI).[6] Secular and left-leaning, it was formed in the 1960s to try to overthrow the Shah of Iran and has been characterized by U.S. reports as attempting to blend several ideologies, including Marxism, feminism, and Islamism, although the organization denies that it ever advocated Marxism. It allied with pro-Khomeini forces during the Islamic revolution and, according to State Department reports, supported the November 1979 takeover of the U.S. Embassy in Tehran, although the group claims that it is the regime that alleged this support in order to discredit the group with the West. The group was driven into exile when it rose up against the Khomeini regime in September 1981. Even though it is an opponent of Tehran, since the late 1980s the State Department has refused contact with the PMOI and its umbrella organization, the National Council of Resistance of Iran (NCRI). It is led by spouses Maryam and Masud Rajavi; Maryam, based in France, is the "President-elect" of the PMOI-led opposition. Masud is the longtime Secretary-General of the PMOI; his whereabouts are unknown.

The "De-Listing" the PMOI

The State Department designated the PMOI as a foreign terrorist organization (FTO) in October 1997[7] and the NCR was named as an alias of the PMOI in the October 1999 re-designation. In August 14, 2003, the State Department designated the NCR offices in the United States an alias of the PMOI, and NCR and the Justice Department closed down those offices.

The PMOI's FTO designation has been widely debated for many years. The State Department's annual reports on international terrorism, including the report for 2011 issued July 31, 2012, asserts that the organization—and not just a radical element of the organization as the group asserts—was responsible for the alleged killing of seven American military personnel and contract advisers to the former Shah during 1973-1976—including the deputy chief of the U.S. Military Mission in Tehran. The report also repeats allegations that the group was responsible for bombings at U.S. government facilities in Tehran in 1972 as a protest of the visit to Iran of then- President Richard Nixon, and bombings of U.S. corporate offices in Iran to protest the visit of Secretary of State Kissinger. The State Department report lists as terrorist acts numerous attacks by the group against regime targets, including major 1981 bombings that killed high ranking officials, and attacks on government facilities in Iran and abroad, and attacks on security officials in Iran. However, the report does not list any attacks by the group that purposely targets civilians—a key distinction that led several experts to argue that the group should not be considered "terrorist." And, the report did not state that the group had, as of mid-2001, fulfilled pledges to end all use of violence inside Iran. There were no reports that the group had resumed those activities. The group's alliance with Saddam Hussein's regime in the 1980s and 1990s has contributed to the U.S. criticism of the organization.

In challenging its FTO decision, the PMOI also asserted that, by retaining the group on the FTO list, the United States was unfairly preventing the PMOI from participating in the opposition movement. The regime accuses the group of involvement in the post-June 2009 presidential election violence, and some of those tried for *mohareb* since February 2010 are members of the organization, according to statements by human rights groups such as Amnesty International.

It also pointed to legal successes in Europe. On January 27, 2009, the European Union (EU) removed the group from its terrorist group list; the group had been so designated by the EU in 2002. In May 2008, a British appeals court determined that the group should no longer be considered a terrorist organization on the grounds that the British government did not provide "any reliable evidence" that the PMOI would "resort to terrorist activities in the future." In June 2003, France arrested about 170 opposition activists, including Maryam Rajavi. She was released and remains based in France, and is frequently received by European parliamentarians and other politicians in Europe. On May 12, 2011, France dropped charges against Mrs. Rajavi and 23 other PMOI activists, saying there was no evidence the PMOI conducted or backed violence against civilians and therefore constitutes resistance, not terrorism.

In July 2008, the PMOI petitioned to the State Department that its designation be revoked on the grounds that it renounced any use of terrorism in 2001. The Department reaffirmed the listing in January 2009 and after a January 2010 review. On July 16, 2010, the Court of Appeals required the State Department to review the listing, ruling that the group had not been given proper opportunity to rebut allegations against it. On February 29, 2012, Secretary Clinton, at a House Foreign Affairs Committee hearing, stated that a "key factor" in the de-listing decision will be the group's compliance with an agreement that its members leave Camp Ashraf, discussed below. That suggests that the department had determined that the group qualifies for removal on the technical issues of involvement in terrorism. However, the group continued to press the issue in court and, in early June 2012, the Appeals Court gave the State Department until October 1, 2012, to decide on the FTO designation, although without prescribing how the Department should decide. On July 6, 2012, two State Department officials reiterated that the relocation from Ashraf (see below) is a key consideration in the FTO decision, and that the group must complete the move, which had stalled.

On September 28, 2012, maintaining there had not been confirmed acts of PMOI terrorism for more than a decade and that it had cooperated with the Ashraf move, the Secretary of State removed the group from the FTO list as well as from the designation as a terrorism supporter under Executive Order 13224. However, State Department officials, in a background briefing that day, said "We do not see the [PMOI] as a viable or democratic opposition movement…They are not part of our picture in terms of the future of Iran." That appeared to counter those who advocate that the United States not only remove the group from the FTO list but also enter an alliance with it.

In an effort to obtain a favorable de-listing decision, supporters of the organization had reportedly been paying several former U.S. officials for panel appearances in which they supported de- listing the group. H.Res. 60, introduced January 26, 2011, "urges" the Secretary of State to remove the PMOI from the FTO list. It attracted nearly 100 co-sponsors.

Camp Ashraf Issue

The issue of group members in Iraq has long been a factor in U.S. policy toward the PMOI. U.S. forces attacked PMOI military installations in Iraq during Operation Iraqi Freedom (March 2003) and negotiated a ceasefire with PMOI military elements in Iraq, requiring the approximately 3,400 PMOI fighters to remain confined to their Ashraf camp near the border with Iran. Its weaponry was placed in storage, guarded first by U.S. and now by Iraqi personnel. Another 200 Ashraf residents took advantage of an arrangement between Iran and the ICRC for them to return to Iran if they disavow further PMOI activities; none is known to have been persecuted since returning.

In July 2004, the United States granted the Ashraf detainees "protected persons" status under the 4th Geneva Convention. However, that designation ended in June 2004 when Iraq formally reassumed full sovereignty from a U.S.-led occupation authority. The U.S.-led, U.N. supported security mandate in Iraq was replaced on January 1, 2009, by a bilateral U.S.-Iraq agreement that limits U.S. flexibility in Iraq. The group long feared that Iraqi control of the camp would lead to the expulsion of the group to Iran. The Iraqi government tried to calm those fears in January 2009 by saying that it would adhere to all international obligations not to do so, but that trust was reduced on July 28, 2009, when Iraq used force to overcome resident resistance to setting up a police post in the camp. Eleven residents of the camp were killed.

The PMOI's fears for Ashraf residents heightened on July 1, 2010, when the Iraqi Security Forces assumed full physical control over Ashraf and the U.S. military post near the camp closed, although U.S. forces in Iraq continued to periodically visit the camp to monitor conditions. On April 2, 2011, with a U.S. military unit overseeing the rotation, the Iraqi government changed the Iraq Security Forces (ISF) brigade that guards Ashraf, triggering PMOI warnings. The U.S. unit departed on April 7, 2011 and clashes between the Iraqi force and camp residents took place on April 8; U.N. High Commissioner for Human Rights Navanethem Pillay largely confirmed PMOI claims that 35 Ashraf residents were killed and that Iraqi forces were at fault. The State Department issued a statement attributing the deaths to the actions of Iraq and its military.[8]

After the clash, Iraqi officials reiterated their commitment to close Ashraf at the end of 2011 (following a full U.S. withdrawal from Iraq), but said such closing would be done in co-operation with the United Nations and other international organizations. On May 16, 2011, the United States offered to help relocate camp residents before Iraq closes it and, in early July 2011, then U.S. Ambassador to Iraq Jim Jeffrey called on the Ashraf residents to disband and seek refugee status elsewhere in Iraq. The U.N. High Commission for Refugees (UNHCR) declared the residents "asylum seekers" and offered to assess each resident in an effort to resettle them elsewhere. The top U.N. envoy in Iraq, Martin Kobler, offered to mediate between the Ashraf residents and the Iraqi government and called on the Iraqi government to postpone its end of 2011 deadline to close the camp. Ambassador Daniel Fried was appointed in November 2011 as the Obama Administration's coordinator on the Ashraf issue.

In late December 2011, Iraqi Prime Minister Nuri al-Maliki extended the deadline for relocation to late as April 2012, and the Iraqi government and the United Nations announced agreement to relocate the residents to former U.S. military base Camp Liberty, near Baghdad International Airport. The PMOI, which had demanded safeguards for their transfer, subsequently announced acceptance of the deal and the move to Camp Liberty (renamed Camp Hurriya). About 2,000 Ashraf residents had relocated as of mid-May 2012, and about 1,200 remained at Ashraf, at which time the PMOI stopped further relocations claiming that conditions at Camp Liberty can't accommodate more residents. In mid-June, and again at the July 6, 2012, briefing by Fried and Coordinator for Counterterorrism Daniel Benjamin (mentioned above), the State Department reiterated that the Secretary of State would link the PMOI's cooperation with the relocation to a decision to take the PMOI from the FTO list, and implied it would not come off the FTO list unless it completes the move. In mid-August 2012, the PMOI resumed the relocation process and largely completed it by September 17, 2012, leaving only a residual group of about 280 PMOI persons at Ashraf to dispose of its property.

The U.N. High Commission on Refugees (UNHCR) is conducting refugee status determinations for all the residents after they relocate. As of early October 2012, several hundred have been interviewed. Two have left Camp Liberty through their links to European countries, not through the resettlement process.

In the aftermath of the April 8, 2011, clashes, H.Res. 231 was introduced, calling on the President to undertake "all necessary and appropriate steps" to ensure the safety and protection of the Ashraf residents. Another bill, H.Res. 332, introduced June 24, 2011, called for a congressional investigation of the incident.

Armed Groups: Pro-Monarchy Radicals

One issue that has arisen in 2010 is that a pro-monarchist armed group in Iran, called Tondar (Thunder)/Kingdom Assembly of Iran is accused of conducting attacks inside Iran. One attack, a bombing of a mosque in Shiraz that took place in April 2008, killed 14 Iranian worshippers, including some children. There are some allegations that Iranians living in California are directing the group's activities in Iran.

Ethnic or Religiously Based Armed Groups

Some armed groups are operating in Iran's border areas, and are generally composed of ethnic or religious minorities. These groups are not known to be cooperating with the mostly Persian members of the Green Movement. One such group is *Jundullah*, composed of Sunni Muslims primarily from the Baluchistan region bordering Pakistan. The region is inhabited by members of the Baluch minority and is far less developed than other parts of Iran. On the grounds that *Jundullah* has attacked civilians in the course of violent attacks in Iran, it was formally placed on the U.S. of Foreign Terrorist Organizations on November 4, 2010. Some saw the designation as an overture toward the Iranian government, while others saw it as a sign that the United States does not support ethnic or sectarian opposition groups that use violence, but only groups that are committed to peaceful protest.

As noted in the State Department terrorism report for 2010, released August 18, 2011, since mid-2006, it has conducted several successful attacks on Iranian security and civilian officials. One of its most widely noted terrorist attacks was a May 2009 bombing of a mosque in Zahedan, which it claimed constituted revenge for the poor treatment of Sunnis in Iran. On October 18, 2009, it claimed responsibility for killing five Revolutionary Guard commanders during a meeting they were holding with local groups in Sistan va Baluchistan Province. The regime claimed a major victory against the group in late February 2010 by announcing the capture of Jundullah's top leader, Abdolmalek Rigi. The regime executed him in June 2010, and the group retaliated in July 2010 with another major bombing in Zahedan, which killed 28 persons, including some Revolutionary Guards. Secretary of State Clinton publicly condemned this bombing. The group is believed responsible for a December 15, 2010, bombing at a mosque in Chahbahar, also in the Baluchistan region, that killed 38 persons.

An armed Kurdish group operating out of Iraq is the Free Life Party, known by its acronym PJAK. Its leader is believed to be Abdul Rahman Hajji Ahmadi, born in 1941, who is a citizen of Germany and lives in that country. Many PJAK members are said to be women, who support the organization's dedication to women's rights. PJAK was designated by the Treasury Department in early February 2009 as a terrorism supporting entity under Executive Order 13224, although the designation statement indicated the decision was based mainly on PJAK's association with the Turkish Kurdish opposition group Kongra Gel, also known as the PKK. The five Kurds executed by Iran's regime in May 2010 were alleged members of PJAK.

In June 2010, Iran conducted some shelling of reputed PJAK bases inside Iraq, reportedly killing some Kurdish civilians. It repeated that activity in July 2011. On September 26, 2011, Turkey's Prime Minister Erdogan said that Iran and Turkey are planning joint operations against the Iraq- based hideouts of these Kurdish opposition groups. Some reports in March 2012 said that PJAK may have reached a ceasefire agreement with the Iranian regime.

Another militant group, the *Ahwazi Arabs*, operates in the largely Arab-inhabited areas of southwest Iran, bordering Iraq. Its activity level appears to have been scant over the past few years.

Iranian-American Interest Groups

Of the more than 1 million Iranian Americans of differing ideologies, a vast majority want to see a change of regime in Tehran. By all accounts, a large number support the Green Movement, although many Iranian Americans are not politically active and focus on their businesses and personal issues. As many as half of all Iranian Americans are based in the Los Angeles area, and they run at least two dozen small-scale radio or television stations that broadcast into Iran. Many of them protest Ahmadinejad's visits to the United Nations General Assembly every September, and many others sport green bracelets showing support for the Green Movement.

National Iranian-American Council, Public Affairs Alliance of Iranian Americans (PAAIA) and Others

Some U.S.-based organizations, such as The National Iranian-American Council (NIAC) and the Public Affairs Alliance of Iranian Americans (PAAIA), are not necessarily seeking change within Iran. The stated mission of NIAC, composed largely of Iranian Americans, is to promote discussion of U.S. policy. The group advocates engagement with Iran, supports easing some U.S. sanctions against Iran, opposes removing the People's Mojahedin (see below) from the U.S. list of terrorist organizations, and has warned that some U.S. experts are seeking to convince the Administration to take military action against Iran. These positions have led some experts and commentators to allege, although without providing evidence, that it is a front for the Iranian regime. On the other hand, NIAC has criticized the regime's human rights abuses.

PAAIA's mission is to discuss issues affecting Iranian Americans, such as discrimination caused by public perceptions of association with terrorism or radical Islam. Some observers believe it has been less active in 2011-2012 than it was in the two previous years, perhaps because of desertions by some who wanted PAAIA to take a stronger stand against the regime in Tehran.

Another U.S.-based group, the International Campaign for Human Rights in Iran, appears supportive of the Green Movement. Believed close to Karrubi and Musavi, it is headed by Hadi Ghaemi. Former CNN anchor Rudi Bahktiar, a relative of the Shah's last prime minister, Shahpour Bakhtiar, has been part of the group. She is an adviser at the Voice of America's Persian News Network (PNN), discussed further below.

OTHER HUMAN RIGHTS PRACTICES

International criticism of Iran's human rights practices predates and transcends the crackdown against the Green Movement. **Table 3**, which discusses the regime's record on a number of human rights issues, is based partly on the latest State Department human rights report (for 2011: May 24, 2012)[9] and on reports from the U.N. Special Rapporteur for human rights in Iran – the latest report of which was issued on September 13, 2012. These reports cite Iran for a wide range of serious abuses—aside from its suppression of political opponents—including unjust executions, politically motivated abductions by security forces, torture, arbitrary arrest and detention, and arrests of women's rights activists. Some human rights groups, including Human Rights Watch, believe that a revised Iranian penal code under consideration in Iran's governing bodies leaves in place much of the legal framework that the regime uses to prosecute dissidents, although the draft revisions might limit child executions.[10]

Criticism of Iran's Record in U.N. Bodies

The post-election crackdown on the Green Movement was a focus of the U.N. four-year review of Iran's human rights record that took place in mid-February 2010 in Geneva. Despite the criticism, on April 29, 2010, Iran acceded to the U.N. Commission on the Status of Women, after earlier dropping its attempt to win a seat on the higher-profile U.N. General Assembly Human Rights Council. Still, on June 10, 2010, Iran was formally questioned by the U.N. Human Rights Council about its record. On November 19, 2010, by a vote of 74-48, with 59 countries abstaining, the General Assembly's "Third Committee" expressed "deep concern" about Iran's forms of punishments and other abuses.

Iran has an official body, the High Council for Human Rights, headed by former Foreign Minister Mohammad Javad Larijani (brother of the Majles speaker and the judiciary head). However, the Council largely acts to defend the government's actions against dissidents to outside bodies and the Iranian public, and does not act as an oversight body urging that Iran meet international standards of human rights practices.

Table 3. Human Rights Practices: General Categories

Group/Issue	Regime Practice/Recent Developments
Ethnic and Religious Breakdown	Persians are about 51% of the population, and Azeris (a Turkic people) are about 24%. Kurds are about 7%-15% of the population, and about 3% are Arab. Of religions, Shiite Muslims are about 90% of the Muslim population and Sunnis are about 10%. About 2% of the population is non-Muslim, including Christians, Zoroastrians (an ancient religion in what is now Iran), Jewish, and Baha'i.
Media Freedoms	Even before the 2009 unrest, Iran's Ministry of Culture and Islamic Guidance had an active program of blocking pro-reform websites and blogs, and had closed hundreds of reformist newspapers, although many have tended to reopen under new names. The State Department human rights report discusses numerous journalists, bloggers, and editors that have been arrested, along with the news organizations they worked for, as well as Iran's efforts to control and monitor Internet usage. The Majles is investigating how and why a blogger, Sattar Beheshti, died in captivity in early November 2012. Iran is setting up a national network that would have a virtual monopoly on Internet service for Iranians.
Labor Restrictions	Independent unions are legal but not allowed in practice. The sole authorized national labor organization is a state-controlled "Workers' House" umbrella.
Women	Women can vote and run in parliamentary and municipal elections, drive, and work outside the home, including owning their own businesses. Nine women are in the Majles. Regime has enforced requirement that women be covered in public, generally with a garment called a chador, but enforcement has reportedly relaxed since 2011. Women do not have inheritance or divorce rights equal to that of men, and their court testimony carries half the weight of a male's.
Religious Freedom	Each year since 1999, the State Department religious freedom report has named Iran as a "Country of Particular Concern" under the International Religious Freedom Act (IRFA). No sanctions have been added under IRFA, on the grounds that Iran is already subject to extensive U.S. sanctions. Continued deterioration in religious freedom noted in the International Religious Freedom report for the second half of 2010, which stated that "Government rhetoric and actions created a threatening atmosphere for nearly all non-Shia religious groups, most notably for Bahais, as well as Sufi Muslims, evangelical Christians, Jews, and Shia groups that do not share the government's official religious views." In late September 2011, a Protestant pastor who was born a Muslim, Youcef Nadarkhani, was sentenced to death for refusing to recant his Christian faith. White House, State Department, and many human rights groups have called for an overturning of the sentence, which was reaffirmed in late February 2012. He was released on September 8, 2012. On February 29, 2012, the House debated but postponed action on H.Res. 556 demanding he be released.
Baha'is	Iran is repeatedly cited for virtually unrelenting repression of the Baha'i community, which Iran's Shiite Muslim clergy views as a heretical sect, which numbers about 300,000-350,000. At least 30 Baha'is remain imprisoned. Several were sentenced to death in February 2010. Seven Baha'i leaders were sentenced to 20 years in August 2010; their sentences were reduced in September 2010 to 10 years but the full sentence was restored on appeal. In the 1990s, several Baha'is were executed for apostasy (Bahman Samandari in 1992; Musa Talibi in 1996; and Ruhollah Ruhani in 1998). Another, Dhabihullah Mahrami, was in custody since 1995 and died of unknown causes in prison in December 2005. Virtually every year, congressional resolutions have condemned Iran's treatment of the Baha'is.
Jews	Along with Christians, a "recognized minority," with one seat in the Majles, the 30,000-member Jewish community (the largest in the Middle East aside from Israel) enjoys somewhat more freedoms than Jewish communities in several other Muslim states. However, in practice the freedom of Iranian Jews to practice their religion is limited, and Iranian Jews remain reluctant to speak out for fear of reprisals. During 1993-1998, Iran

Table 3. (Continued)

Group/Issue	Regime Practice/Recent Developments
	executed five Jews allegedly spying for Israel. In June 1999, Iran arrested 13 Jews (mostly teachers, shopkeepers, and butchers) from the Shiraz area that it said were part of an "espionage ring" for Israel. After a 2000 trial, 10 of the Jews and 2 Muslim accomplices were convicted and given sentences ranging from 4 to 13 years. An appeals panel reduced the sentences, and all were released by April 2003. On November 17, 2008, Iran hanged businessman Ali Ashtari (a Muslim), for providing information on Iran's nuclear program to Israel.
Azeris	Azeris are one-quarter of the population and are mostly well integrated into government and society, but many Azeris complain of ethnic and linguistic discrimination. Each year, there are arrests of Azeri students and cultural activists who press for their right to celebrate their culture and history. The government generally accuses these activists of promoting revolution or separatism.
Kurds	There are about 5 million-11 million Kurds in Iran. The Kurdish language is not banned, but schools do not teach it and Kurdish political organizations, activists, and media outlets are routinely scrutinized, harassed, and closed down for supporting greater Kurdish autonomy. Several Kurdish oppositionists have been executed since 2010.
Arabs	Ethnic Arabs are prominent in southwestern Iran, particularly Khuzestan Province. The 2 million to 4 million Arabs in Iran encounter oppression and discrimination, including torture and a prohibition on speaking or studying Arabic.
Human Trafficking	The June 19, 2012 (latest), State Department "Trafficking in Persons" report, for the seventh consecutive year, places Iran in Tier 3 (worst level) for failing to take significant action to prevent trafficking in persons. Iranian women, boys, and girls are trafficked for sexual exploitation in Iran as well to Pakistan, the Persian Gulf, and Europe, possibly with the involvement of religious leaders and immigration officials.
Executions Policy	Human rights groups say executions have increased sharply since the dispute over the June 2009 election. The State Department human rights report for 2011 said there were between 275 and 700 executions during 2011—disparate figures cited by various human rights organizations. The numbers included several persons for crimes committed when they were minors. Iran is a party to the International Covenant on Civil and Political Rights and the Convention on the Rights of the Child, and is obligated to cease the executions of minors. As noted above, a revised penal code under consideration would prevent child executions in many, but not all, circumstances.
Stonings	In 2002, the head of Iran's judiciary issued a ban on stoning. However, Iranian officials later called that directive "advisory" and could be ignored by individual judges. On December 2, 2008, Iran confirmed the stoning deaths of two men in Mashhad who were convicted of adultery. A sentence of stoning against a 45-year-old woman (Sakineh Ashtiani) convicted of adultery and assisting in the murder of her husband was set aside for further review in July 2010. An Iranian parliamentarian said on January 17, 2011, the stoning sentence was dropped but she would serve 10 years in prison.
Arrests of Dual Nationals and Foreign Nationals/ Robert Levinson/ the American Hikers	An Iranian American journalist, Roxanna Saberi, was arrested in January 2009 allegedly because her press credentials had expired; was charged on April 9, 2009, with espionage for possessing an Iranian military document. Sentenced to eight years in jail, she was released on appeal on May 12, 2009, and left Iran. U.S. national, former FBI agent Robert Levinson, remains missing after a visit in 2005 to Kish Island. In December 2011, his family released a one-year old taped statement by him and appealed for help in obtaining his release, although Iran said it does not know where he is. *Hikers*. Three American hikers were arrested in August 2009 after crossing into Iran, possibly mistakenly, from a hike in northern Iraq. On September 15, 2010, Sara Shourd was released on $500,000 bail, an arrangement brokered by Oman. Her fiancé, Shane Bauer, and Josh Fattal, were released on September 21, 2011, on similar terms.

Sources: Most recent State Department reports on human rights (May 24, 2012, trafficking in persons (June 19, 2012), and on religious freedom (September 13, 2011). http://www.state.gov; U.N. Special Rapporteur report dated September 23, 2011 (U.N. document no. A/66/374).

Special U.N. Rapporteur Reestablished

On February 28, 2011, in remarks at the U.N. Human Rights Council in Geneva, Secretary Clinton said the United States is working with Sweden and other countries to reconstitute a Special Rapporteur to report on Iranian human rights abuses. Such a mission existed during the from 1988-2002, but Iran tended to offer little, if any, cooperation with the various Rapporteurs who investigated the issue during that time. On March 24, 2011, the U.N. Human Rights Council voted, 22 to 7, to reestablish a Special Rapporteur for Iran's human rights situation. On June 17, 2011, former Maldives Foreign Minister Ahmad Shaheed was appointed to this role, but he has said Iran has not, to date, provided him the requested cooperation such as permitting him to conduct fact-finding visits to Iran. The Rapporteur issued his first report on September 23, 2011 (U.N. Document Number A/66/374: "The Situation of Human Rights in the Islamic Republic of Iran"), and subsequent reports on March 6, 2012 (A/HRC/19/66) and September 13, 2012 (A/67/369). The reports cite many of the same abuses as do the State Department reports mentioned above.

On November 21, 2011, the U.N. General Assembly's Third Committee, by a vote of 86-32, with 59 abstentions, approved a resolution asserting that Iran must cooperate with the efforts of the Special Rapporteur to assess the human rights situation in Iran. The full Assembly approved the resolution on December 19, 2011, by a vote of 89-30 with 64 abstentions.

IRAN'S STRATEGIC CAPABILITIES AND WEAPONS OF MASS DESTRUCTION PROGRAMS

The Obama Administration views Iran as one of the key national security challenges facing the United States – an assessment based largely on suspicions about Iran's nuclear and missile programs and its intent and ability to counter U.S. objectives in the region. A nuclear armed Iran, in the view of U.S. and Persian Gulf state officials, would be more assertive than it now is in trying to influence the foreign and energy policies of the Persian Gulf states and in supporting countries and movements that oppose U.S. interests and allies. Iran would likely conclude that the United States would hesitate to take military action against—or undertake any action to try to change the regime of—a nuclear-armed Iran. A nuclear-armed Iran would also likely cause other countries in the region to try to acquire a countervailing nuclear capability—stimulating a nuclear arms race in one of the world's most volatile regions. Israel views an Iranian nuclear weapon as a threat to its existence.

Others see Iran's foreign policy as primarily defensive. Some believe Iran's core national security goals are to protect itself from foreign, primarily U.S., interference or attack; to prevent any efforts to cut off its ability to export oil; and to exert regional influence that Iran believes is commensurate with its size and concept of nationhood.

Conventional Military/Revolutionary Guard/Qods Force

Iran's armed forces are extensive but they are widely considered relatively combat ineffective in a head-on confrontation against a well-trained, sophisticated military such as that of the United States or even a major regional power such as Turkey. Iran is believed to largely lack the logistical ability to deploy ground forces much beyond its borders. However, a 2012 Defense Department report, required by P.L. 111-84, reported growing lethality and survivability of Iran's ballistic and cruise missiles, suggesting the Defense Department assesses a higher level of conventional threat from Iran as compared to a similar DOD report in 2010.[11] The 2012 assessment raises the question of whether Iran possesses the capability to close the strategic Strait of Hormuz, where about one-third of all seaborne traded oil flows—an issue that has gained greater urgency in 2012 as Iranian leaders and commanders talked openly of trying to do so as a retaliation for international sanctions against Iran's oil exports. The Iranian armed forces are sufficiently effective to deter or fend off any threats, should they emerge, from Iran's weaker neighbors such as post-war Iraq, Turkmenistan, Azerbaijan, and Afghanistan.

Table 4. Iran's Conventional Military Arsenal

Military Personnel: 460,000+. Regular ground force is about 220,000, Revolutionary Guard Corps (IRGC) ground force is about 130,000. Remainder are regular and IRGC navy (18,000 and 20,000 personnel respectively) and Air Forces (52,000 regular Air Force personnel and 5,000 Guard Air Force personnel.) About 12,000 air defense.
Security Forces: About 40,000-60,000 law enforcement forces on duty, with another 600,000 Basij security/paramilitary forces available for combat or internal security missions.
Tanks: 1,800+ Includes 480 Russian-made T-72.
Ships: 100+ (IRGC and regular Navy) Includes 4 Corvette; 18 IRGC-controlled Chinese-made patrol boats, several hundred small boats.) Also has 3 Kilo subs (reg. Navy controlled). 2012 DOD report says Iran may have acquired additional ships and submarines over the past two years, but does not stipulate a supplier, if any.
Midget Subs: Iran has been long said to possess several small subs, possibly purchased assembled or in kit form from North Korea. Iran claimed on November 29, 2007, to have produced a new small sub equipped with sonar-evading technology, and it claimed to deploy four Iranian-made "Ghadir class" subs to the Red Sea in June 2011.
Surface-to-Air Missiles (SAMs): 150+ I-Hawk plus possibly some Stinger
Combat Aircraft: 330+ Includes 25 MiG-29 and 30 Su-24. Still dependent on U.S. F-4's, F-5's and F-14 bought during Shah's era.
Anti-aircraft Missile Systems: Russia delivered to Iran (January 2007) 30 anti-aircraft missile systems (Tor M1), worth over $1 billion. In September 2006, Ukraine agreed to sell Iran the Kolchuga radar system that can improve Iran's detection of combat aircraft. In December 2007, Russia agreed to sell the highly capable S-300 (also known as SA-20 "Gargoyle") air defense system, which would greatly enhance Iran's air defense capability. The value of the deal is estimated at $800 million. The system is a ground-to-air missile whose sale to Iran would, according to most experts, not technically violate the provisions of U.N. Resolution 1929, because the system is not covered in the "U.N. Registry on Conventional Arms. However, on September 22, 2010, Russian President Medvedev signed a decree banning the supply of the system to Iran, asserting that its provision to Iran is banned by Resolution 1929. In November, Iran claimed to have deployed its own version (Mersad) of the Russian S-200 air defense system and said in September 2012 that it has completed 30% of a missile defense network similar to the S-300. In August 2011, Iran sued Russia at the International Court of Justice for non-delivery of the system.
Defense Budget: About 3% of GDP

Sources: IISS Military Balance—Section on Middle East and North Africa, and various press reports; April 2010 and April 2012 DOD reports on military power of Iran," cited earlier.

Organizationally, Iran's armed forces are divided to perform functions appropriate to their roles in Iran. The Islamic Revolutionary Guard Corps (IRGC, known in Persian as the *Sepah-e-Pasdaran Enghelab Islami*)[12] controls the Basij (Mobilization of the Oppressed) volunteer militia that has been the main instrument to repress Green Movement protests in Iran. The IRGC and the regular military (*Artesh*) report to a joint headquarters, headed by Dr. Hassan Firuzabadi. The *Artesh* has no role in internal security and is deployed mainly at bases outside major cities.

The IRGC Navy and regular Navy (Islamic Republic of Iran Navy, IRIN) are distinct forces; the IRIN has responsibility for the Gulf of Oman, whereas the IRGC Navy has responsibility for the closer-in Persian Gulf and Strait of Hormuz. The regular Air Force controls most of Iran's combat aircraft, whereas the IRGC Air Force has come to focus primarily on developing Iran's ballistic missile capabilities.

Iran's armed forces have few formal relationships with foreign militaries, but Iran and India have a "strategic dialogue" and some Iranian naval officers reportedly have undergone some training in India. Most of Iran's other military-to-military relationships, such as with Russia, Ukraine, Belarus, North Korea, and a few others, generally center on Iranian arms purchases or upgrades, although such activity is now banned by U.N. Resolution 1929 of June 2010. In early September 2012, Iran and North Korea signed an agreement to cooperate on science and technology, raising concerns about potential additional North Korean support to Iran's nuclear program.

Nuclear Program and Related International Diplomacy

International attention to Iran's nuclear program intensified in late 2002, when Iran confirmed PMOI allegations that Iran was building two facilities that could potentially be used to produce fissile material useful for a nuclear weapon: a uranium enrichment facility at Natanz and a heavy water production plant at Arak,[13]considered ideal for the production of plutonium. The United States and its partners state that they accept Iran's right to pursue peaceful uses of nuclear energy, but that Iran must verifiably demonstrate that its nuclear program is for only those purposes. Iran has been producing low-enriched uranium (3.5%-5%, suitable only for electricity production) since 2002 and more recently has been enriching to 20% (which Iran says is being used to produce medical isotopes), as discussed below. To construct an actual nuclear weapon, Iran would have to produce highly enriched uranium (90%+) and master the complicated capability to trigger a nuclear detonation.

Iran's Nuclear Intentions[14]

U.S. officials have stated on several occasions in 2012 that the United States believes Iran has not, to date, made a decision to construct a nuclear weapon. However, International Atomic Energy Agency (IAEA) reports indicate that Iran has not satisfactorily addressed IAEA information that Iran might have a nuclear *weapons* program.[15] Several pre-2011 IAEA reports describe Iranian documents that show a possible involvement of Iran's military in the program. This issue garnered heightened attention after the IAEA released its November 8, 2011, report that contained an extensive annex laying out the IAEA's information on Iran's apparent efforts to acquire the knowledge required to weaponize highly enriched uranium, and on some possible facilities Iran had constructed that could be used for that effort. The annex discussed the IAEA's sources, purported foreign scientific assistance to the experimentation, and Iran's management structure for a weapons program. Based on the report, on November 18, 2011, the IAEA Board of Governors adopted a resolution expressing "deep and increasing concern" about Iran's nuclear program. The vote was 32 in favor, 2 against (Cuba, Ecuador), and 1 abstention (Indonesia).

Table 5. The Revolutionary Guard

The IRGC is generally loyal to Iran's hardliners politically and is clearly more politically influential than is Iran's regular military, which is numerically larger, but was held over from the Shah's era. IRGC influence has grown sharply as the regime has relied on it to suppress dissent to the point where Secretary of State Clinton sees it as wielding preponderant influence. As described in a 2009 Rand Corporation study," Founded by a decree from Ayatollah Khomeini shortly after the victory of the 1978-1979 Islamic Revolution, Iran's Islamic Revolutionary Guards Corps (IRGC) has evolved well beyond its original foundations as an ideological guard for the nascent revolutionary regime. Today the IRGC functions as an expansive socio-political-economic conglomerate whose influence extends into virtually every corner of Iranian political life and society. Bound together by the shared experience of war and the socialization of military service, the Pasdaran have articulated a populist, authoritarian, and assertive vision for the Islamic Republic of Iran that they maintain is a more faithful reflection of the revolution's early ideals. The IRGC's presence is particularly powerful in Iran's highly factionalized political system, in which [many senior figures] hail from the ranks of the IRGC. Outside the political realm, the IRGC oversees a robust apparatus of media resources, training activities, education programs designed to bolster loyalty to the regime, prepare the citizenry for homeland defense, and burnish its own institutional credibility vis-à-vis other factional actors."
Through its Qods (Jerusalem) Force, the IRGC has a foreign policy role in exerting influence throughout the region by supporting pro-Iranian movements, as discussed further below. The Qods Force numbers approximately 10,000- 15,000 personnel who provide advice, support, and arrange weapons deliveries to pro-Iranian factions in Lebanon, Iraq, Persian Gulf states, Gaza/West Bank, Afghanistan, and Central Asia. IRGC leaders have confirmed it is operating in Syria to assist the regime of Bashar al-Assad against an armed uprising. It also operates a worldwide intelligence network to give Iran possible terrorist option and to assist in procurement of WMD-related technology. The Qods Force commander, Brigadier General Qassem Soleimani, is said to have his own independent channel to Supreme Leader Khamene'i, bypassing the IRGC and Joint Staff command structure. The Qods Force commander during 1988- 1995 was Brigadier General
Ahmad Vahidi, confirmed as defense minister on September 3, 2009. He led the Qods Force when it allegedly assisted two bombings of Israeli and Jewish targets in Buenos Aires (he is wanted by Interpol for a role in the 1994 bombing there); recruited Saudi Hezbollah activists later accused of the June 1996 Khobar Towers bombing; and assassinated Iranian dissident leaders in Europe in the early 1990s.
IRGC leadership developments are significant because of the political influence of the IRGC. On September 2, 2007, Khamene'i named Mohammad Ali Jafari as commander in chief of the Guard; Jafari is considered a hardliner against political dissent and increasingly at odds with Ahmadinejad in the context of the Ahmadinejad-Khamene'i power struggle. The *Basij* reports to the IRGC commander in chief; its leader is Brigadier General Mohammad Reza Naqdi. It operates from thousands of positions in Iran's institutions. Command reshuffles in July 2008 integrated the Basij more closely with provincially based IRGC units and increased the Basij role in internal security. In November 2009, the regime gave the IRGC's intelligence units greater authority, perhaps surpassing those of the Ministry of Intelligence, in monitoring dissent. The IRGC Navy has responsibility to patrol the Strait of Hormuz and the regular Navy has responsibility for the broader Arabian Sea and Gulf of Oman (deeper waters further off the coast).
As noted, the IRGC is also increasingly involved in Iran's economy, acting through a network of contracting businesses it has set up, most notably *Ghorb* (also called *Khatem ol-Anbiya*, Persian for

"Seal of the Prophet"). Active duty IRGC senior commanders reportedly serve on Ghorb's board of directors and its commander, Rostam Ghasemi, became oil minister in August 2011. In September 2009, the Guard bought a 50% stake in Iran Telecommunication Company at a cost of $7.8 billion. In the past five years, Guard affiliated firms have won 750 oil and gas and construction contracts, and the Guard has its own civilian port facilities. However, *Ghorb* pulled out of a contract to develop part of the large South Pars gas field in July 2010, citing the impact of expanded U.S. and international sanctions, which might have caused foreign partner firms to refuse to cooperate with *Ghorb*.
On October 21, 2007, the Treasury Department designated several IRGC companies as proliferation entities under Executive Order 13382. Also that day, the IRGC as a whole, the Ministry of Defense, several IRGC commanders, and several Iranian banks were sanctioned under that same executive order. Simultaneously, the Qods Force was named as a terrorism supporting entity under Executive Order 13224. These orders freeze the U.S.-based assets and prevent U.S. transactions with the named entities, but these entities are believed to have virtually no U.S.-based assets. On June 9, 2011, the IRGC and Basij were named as human rights abusers under Executive Order 13553, with the same penalties as the above Executive Orders.

Sources: Frederic Wehrey et al. "The Rise of the Pasdaran." Rand Corporation. 2009. Katzman, Kenneth. "The Warriors of Islam: Iran's Revolutionary Guard." Westview Press, 1993; Dept. of the Treasury.

After repeatedly refusing to discuss the IAEA information, in January 2012 Iran agreed to host an IAEA team to discuss the allegations during January 29-31, 2012. That visit, as well as a subsequent IAEA visit during February 20-21, 2012, did not satisfy the IAEA on the question of Iran's nuclear *weapons* research. Iran did not allow the IAEA team, in either trip, to visit the Parchin military base where the IAEA suspects research on nuclear explosive technology may have taken place. The site was inspected twice in 2005. IAEA Director Yukiya Amano, following an unexpected visit to Iran on May 21, 2012, announced an "agreement in principle" under which Iran would clear up outstanding questions about its suspected nuclear weapons design work. However, amid IAEA accusations – repeated in November 2012 - that Iran may have cleaned up parts of the Parchin facility, no pact was finalized. That precipitated another IAEA Board of Governors resolution criticizing Iran for non-cooperation; it was adopted on September 13, 2012, with 31 countries in favor, Cuba against, and Ecuador, Tunisia, and Egypt abstaining. In early November 2012, Iran and the IAEA agreed to hold further talks in mid-December 2012 to try to finalize the agreement under discussion.

Iran's Position and Counter-Arguments

Iranian leaders deny they are trying to achieve a nuclear weapons capability and they assert that IAEA information demonstrates little more than that some of its scientists may have performed nuclear weapons calculations on computers. They assert that Iran's nuclear program is for medical uses and electricity generation, given finite oil and gas resources, and that enrichment is its "right" as a party to the 1968 Nuclear Non-Proliferation Treaty.[16] They cite studies, including an analysis by the National Academy of Sciences, that Iran might have negligible exports of oil by 2015.[17] U.S. officials have said that Iran's gas resources make nuclear energy unnecessary.

Iran professes that WMD is inconsistent with its ideology. In 2003, the Supreme Leader Khamene'i issued a formal pronouncement (*fatwas*) that nuclear weapons are un-Islamic. On February 22, 2012, he expanded on that concept in a speech saying that the production of and use of a nuclear weapon is prohibited as a "great sin," and that stockpiling such weapons is "futile, expensive, and harmful."[18] He repeated this formulation in an August 30, 2012, speech to the Non-Aligned Movement summit meeting in Tehran.

Iran's assertions of a purely peaceful program have been met with widespread skepticism, not only because of the activities discussed above but also because Iran's governing factions appear to perceive a nuclear weapons capability as a means of ending Iran's perceived historic vulnerability to invasion and domination by great powers, and as a symbol of Iran as a major nation. Others believe a nuclear weapon represents the instrument with which Iran intends to intimidate its neighbors and dominate the Persian Gulf region. Still others believe regime leaders see a nuclear weapon as insurance that domestic or international opponents will end perceived attempts to displace the regime. There are also fears Iran might transfer WMD to extremist groups or countries.

Aside from the issue about the cost international sanctions are imposing on Iran, some Iranian strategists appear to agree with U.S. assertions that a nuclear weapon will not deliver Iran absolute security, but will instead make Iran less secure. According to this view, moving toward a nuclear weapons capability will bring Iran further sanctions, military containment, U.S. attempted interference in Iran, and efforts by neighbors to develop countervailing capabilities. Some Green Movement leaders have positions on the nuclear issue similar to those of regime leaders, but many Green Movement factions see the nuclear program as an impediment to eventual reintegration with the West.

Nuclear Weapons Time Frame Estimates

If Iran were to pursue a nuclear weapon, estimates differ as to when Iran might achieve that capability. Secretary of Defense Panetta said on January 29, 2012, that Iran could produce a nuclear weapon within about one year of a decision to do so, and a delivery vehicle for that weapon one to two years after that. Some outside groups, such as the Institute for Science and International Security, in a study released in early October 2012, make similar assessments. These estimates take into account technical difficulties, possibly caused by Western activities and international sanctions, as well as reported covert action that might have delayed a nuclear-armed Iran. National Security Adviser Donilon stated in his Brookings Institution speech on November 22, 2011, that sanctions and other difficulties "have succeeded in slowing [Iran's] nuclear program." Among these difficulties is the effect of a deliberate computer virus (Stuxnet) in September-October 2010 that appeared to target Iranian nuclear facility computers by altering their spin rate,[19] although IAEA reports since 2011 indicate that Iran overcame these effects.

A separate but related issue is the ability of the United States and IAEA to detect an all-out effort by Iran to develop an actual nuclear weapon. Defense Secretary Panetta said on September 11, 2012, that the United States has "pretty good intelligence" on Iran's nuclear program. The crucial role of the IAEA in helping determine whether Iran is trying to construct a weapon explains why many experts are concerned that Iran might reduce cooperation with the IAEA. Those concerns have been raised since September 18, 2012, when its nuclear chief, Abbas Davani, accused unspecified attackers of trying to sabotage the electricity supply to Iran's nuclear facilities—an accusation that came at the time of an IAEA visit to the country. In addition, some Iranian parliamentarians and officials have openly threatened to reduce cooperation with the IAEA.[20]

Status of Enrichment

Sparking further concerns among several governments is the steady progress of Iran's enrichment program, and the August 30, 2012, IAEA report indicates that enrichment has continued. Thus far, according to that report, Iran has a stockpile of enough low-enriched (3.5%—5% uranium (15,100 lbs+) to produce as many as five nuclear weapons, if it were to enrich that stockpile to weapons grade. It has a stockpile of about 417 lbs of 20% enriched uranium—up about 60% from the amounts reported in the previously quarterly IAEA report—but still not enough to produce a nuclear weapon if it were using only that stockpile.[21] However, according to the August 30, 2012, IAEA report, some of the 20% enriched uranium has been used to fabricate fuel rods that could not be used to produce weapons-grade uranium. To some extent, this has undercut the Israeli and other arguments that Iran is accumulating 20% enriched uranium for the purpose of developing a nuclear weapon.

Some of the enrichment to 20% is taking place at the heavily fortified Fordow site that Iran admitted in September 2009 (after discovery by Western intelligence) that it had developed. The August 30, 2012, IAEA report said Iran had added about 1,000 centrifuges at that site, although they are not yet operational. The IAEA report added that Iran has had difficulty with its newer generation centrifuges (IR-2M and IR-4) at Fordow and at the main Natanz enrichment site. In late October 2012, press reports said that Iran had completed installing the last of the 3,000 centrifuges planned for at Fordow, although only about half the centrifuges are operating. In late August 2011, the head of Iran's atomic energy agency said Iran would produce more 20% enriched uranium than it needs for the medical reactor, increasing concern about Iran's intentions.

The IAEA reports maintain that there is no evidence that Iran has diverted any nuclear material (for a nuclear weapons program). The February 25, 2011, IAEA report has annexes listing Iran's declared nuclear sites as well as a summary of all the NPT obligations Iran is not meeting.[22]

Bushehr Reactor

U.S. officials have generally been less concerned with Russia's work, under a January 1995 contract, on an $800 million nuclear power plant at Bushehr. Russia insisted that Iran sign an agreement under which Russia would reprocess the plant's spent nuclear material; that agreement was signed on February 28, 2005. The plant was expected to become operational in 2007, but Russia appeared to delay opening it to pressure Iran on the broader nuclear issue. The plant was inaugurated on August 21, 2010, and fueling was completed by October 25, 2010. It began limited operations on May 8, 2011, and was linked to Iran's power grid in September 2011. It was reported by Iran as operational as of September 3, 2012, but it is still mostly operated by Russia and a Russian energy official said the plant would not be fully operational and turned over to full Iranian control until early 2013. As part of this work, Russia trained 1,500 Iranian nuclear engineers.

Early International Diplomatic Efforts to Address Iran's Nuclear Program

The international response to Iran's nuclear program has been longstanding, slowly evolving into a growing global consensus to apply substantial pressure on Iran—coupled with diplomacy and selected incentives—to limit its program. In 2003, France, Britain, and Germany (the "EU-3") opened a separate diplomatic track to curb Iran's program, initially having some success. On October 21, 2003, Iran pledged, in return for peaceful nuclear technology, to (1) fully disclose its past nuclear activities, (2) sign and ratify the "Additional Protocol" to the NPT (allowing for enhanced inspections), and (3) suspend uranium enrichment activities. Iran signed the Additional Protocol on December 18, 2003, although the *Majles* has not ratified it. Iran discontinued abiding by the Protocol after the IAEA reports of November 10, 2003, and February 24, 2004, stated that Iran had violated its NPT reporting obligations over an 18-year period.

In the face of the U.S. threat to push for Security Council action, the EU-3 and Iran reached a more specific November 14, 2004, "Paris Agreement," committing Iran to suspend uranium enrichment (which it did as of November 22, 2004) in exchange for renewed trade talks and other aid.[23] The Bush Administration did not openly support the track until March 11, 2005, when it announced it would drop U.S. objections to Iran applying to join the World Trade Organization (it applied in May 2005) and to selling civilian aircraft parts to Iran. The Bush Administration did not participate directly in the talks.

The Paris Agreement broke down just after Ahmadinejad's election; Iran rejected as insufficient an EU-3 offer to permanently resolve the issue by assisting Iran with peaceful uses of nuclear energy and provide limited security guarantees, in exchange for several Iranian steps : (1) permanently ending uranium enrichment; (2) dismantling the Arak heavy-water reactor;[24] (3) allowing no-notice nuclear inspections; and (4) pledging not to leave the NPT (it has a legal exit clause). On August 8, 2005, Iran broke the IAEA seals and began uranium "conversion" (one step before enrichment) at its Esfahan facility. On September 24, 2005, the IAEA Board declared Iran in non-compliance with the NPT and decided to refer the issue to the Security Council,[25] but no time frame was set for the referral. After Iran resumed enrichment activities, on February 4, 2006, the IAEA board voted 27-3[26] to refer the case to the Security Council. On March 29, 2006, the Council agreed on a presidency "statement" setting a 30-day time limit (April 28, 2006) for ceasing enrichment.[27]

Taking a multilateral approach, the George W. Bush Administration offered on May 31, 2006, to join the nuclear talks with Iran if Iran first suspends its uranium enrichment. Such talks would center on a package of incentives and possible sanctions—formally agreed on June 1, 2006—by a newly formed group of nations, the so-called "Permanent Five Plus 1" (P5+1: United States, Russia, China, France, Britain, and Germany). EU foreign policy chief Javier Solana – acting as representative of the P5+1 - formally presented the P5+1 offer to Iran on June 6, 2006. The incentive package—which includes accepting Iran into the World Trade Organization and Persian Gulf security frameworks, easing sanctions, and guarantees of nuclear fuel and light-water reactor technology, and is Annex I to Resolution 1747, and the offer remains on the table, according to U.S. and EU diplomats. Sanctions threatened[28] - such as a ban on technology and arms sales to Iran - have mostly been imposed subsequently.

First Set of U.N. Security Council Resolutions Adopted

Iran did not immediately respond to the offer. In response, the U.N. Security Council began to impose sanctions on Iran in an effort to shift Iran's calculations toward compromise.

- *Resolution 1696*. On July 31, 2006, the Security Council voted 14-1 (Qatar voting no) for U.N. Security Council Resolution 1696, giving Iran until August 31, 2006, to fulfill the long-standing IAEA nuclear demands (enrichment suspension, etc.).

Purportedly in deference to Russia and China, it was passed under Article 40 of the U.N. Charter, which makes compliance mandatory, but not under Article 41, which refers to economic sanctions, or Article 42, which would authorize military action. It called on U.N. member states not to sell Iran WMD-useful technology.

- *Resolution 1737.* After Iran refused a proposal to temporarily suspend enrichment, the Security Council adopted U.N. Security Council Resolution 1737 unanimously on December 23, 2006, under Chapter 7, Article 41 of the U.N. Charter. It demanded enrichment suspension by February 21, 2007 and prohibits sale to Iran—or financing of such sale—of technology that could contribute to Iran's uranium enrichment or heavy-water reprocessing activities. It also required U.N. member states to freeze the financial assets of several named Iranian nuclear and missile firms and related persons. In deference to Russia, the Resolution exempted the Bushehr reactor.
- *Resolution 1747.* With no Iranian compliance, on March 24, 2007, after only three weeks of P5+1 negotiations, Resolution 1747 was adopted unanimously. It demanded Iran suspend enrichment by May 24, 2007, added entities to those sanctioned by Resolution 1737, and:
  - banned arms transfers by Iran, a provision targeted at Iran's alleged arms supplies to Lebanese Hezbollah and to Shiite militias in Iraq;
  - required all countries to report to the United Nations when sanctioned Iranian persons travel to their territories; and
  - called for (but did not require) countries to avoid selling arms or dual use items to Iran and for countries and international financial institutions to avoid any new lending or grants to Iran. The Resolution specifically exempted loans for humanitarian purposes, thereby not applying to World Bank loans.
- *Resolution 1803 and Additional Incentives for Iran.* With no Iranian compliance forthcoming, after several months of negotiations, Resolution 1803 was adopted by a vote of 14-0 (Indonesia abstaining) on March 3, 2008. It added 12 more entities to those sanctioned, and:
  - banned virtually all sales of dual use items to Iran, citing equipment listed as dual use in various proliferation conventions and documents;
  - authorized, but did not require, inspections of shipments by Iran Air Cargo and Islamic Republic of Iran Shipping Line, if such shipments are suspected of containing banned WMD-related goods;
  - imposed a firm travel ban on five Iranians named in Annex II to the Resolution and requires reports on travel by 13 other named individuals; and
  - stated the willingness of the P5+1 to consider additional incentives to resolve the Iranian nuclear issue through negotiation "beyond those of June 2006."

The Bush Administration agreed to expand the June 2006 incentive package at a meeting in London on May 2, 2008, offering political and enhanced energy cooperation with Iran. EU envoy Solana presented the package (which included a signature by Secretary of State Rice) on June 14,

2008, but Iran was non-committal. (The text of the enhanced incentive offer to Iran is contained in an Annex to Resolution 1929.) Iran did not accept the enhanced package of incentives but, in July 2008, it indicated it might be ready to first accept a six week "freeze for freeze:" the P5+1 would freeze further sanctions efforts and Iran would freeze any expansion of uranium enrichment. To try to take advantage of this opening, the Bush Administration sent Under Secretary of State for Political Affairs William Burns to join Solana and the other P5+1 representatives at a meeting in Geneva on July 19, 2008. Iran did not accept the "freeze for freeze" by an extended deadline of August 2, 2008.

- *Resolution 1835*. The August 2008 crisis between Russia and Georgia contributed to Russia's opposing new U.N. sanctions on Iran. In an effort to demonstrate to Iran continued unity, the Council adopted Resolution 1835 (September 27, 2008), demanding compliance with previous resolutions but not adding new sanctions.

Developments under the Obama Administration

After President Obama was inaugurated, the P5+1 met in Germany on February 4, 2009, seeking to incorporate into the joint approach the U.S. Administration's commitment to direct U.S. engagement with Iran into the U.N. sanctions and negotiating framework.[29] On April 8, 2009, then Under Secretary William Burns told the other members of the P5+1 that a U.S. diplomat would attend all of the group's meetings with Iran. The P5+1 did not materially alter its approach because of the unrest in Iran that erupted after that election, but a July 9, 2009, G-8 summit statement, which included Russian concurrence, mentioned that the P5+1 expected Iran to offer constructive proposals by late September 2009 or face "crippling sanctions." On September 9, 2009, Iran distributed its proposals to settle the nuclear issue to P5+1 representatives in Iran, which the P5+1 considered vague but still a sufficient basis to meet with Iran on October 1, 2009.

October 1, 2009, Agreement on Reprocessing Iran's Enriched Uranium

In light of September 25, 2009, revelations about the previously unreported Iranian nuclear site, little progress was expected at the October 1, 2009 meeting in Geneva. However, the meeting resulted in an apparent breakthrough in the form of a tentative agreement to allow Russia and France, subject to technical talks to begin by mid-October, to reprocess 2,600 pounds (about 75% of Iran's low-enriched uranium) for medical use. There was also agreement for the Fordow facility to be inspected, which happened during October 25-29, 2009. At the session, Burns, representing the United States, also met privately with Iranian negotiator Sayed Jallili.

The technical talks were held October 19-21, 2009, at IAEA headquarters in Vienna, Austria, and chaired on the U.S. side by Deputy Energy Secretary Daniel Poneman. A draft agreement was approved by the P5+1 countries and the IAEA. Despite Ahmadinejad's comments in early February 2010 that he "did not have a problem" with the arrangement, political opposition from hardliners inside Iran—purportedly including the Supreme Leader himself—caused Iran to refuse to finalize the agreement. Instead, Iran floated counter-proposals to ship its enriched uranium to France and Russia in increments, to ship the uranium to Turkey, or to reprocess the uranium in Iran itself. All of Iran's counter-proposals were deemed insufficiently specific or responsive. Iran also rebuffed a specific U.S. proposal in January 2010 to allow it to buy on the open market isotopes for its medical reactor, although this proposal remains active.

May 2010 Iran-Brazil-Turkey Uranium Exchange Deal ("Tehran Declaration")

As international discussions of new sanctions accelerated in April 2010, Brazil and Turkey negotiated with Iran to revive the October 1, 2009, arrangement. On May 17, 2010, with the president of Brazil and prime minister of Turkey in Tehran, the three signed an arrangement for Iran to send 2,600 pounds of uranium to Turkey, which would be exchanged for medically useful reprocessed uranium along the lines discussed in October 2009.[30] As required by the agreement, Iran forwarded to the IAEA a formal letter accepting the agreement terms. Even though some assert that the Obama Administration quietly supported the Brazil-Turkey initiative, the Obama Administration did not accept the Tehran Declaration, asserting, primarily, that the arrangement did not address Iran's enrichment to the 20% level.

Resolution 1929 and EU Follow-Up

On May 18, 2010, one day after the signing of the Tehran Declaration, Secretary of State Clinton announced that the P5+1 had reached agreement on a new sanctions resolution. The resolution was designed to attract support from Russia and China, which believe sanctions might threaten their own interests in Iran, while also giving U.S. allies authority to take substantial new measures against Iran. Simultaneous with Russian agreement on the draft, several Russian entities, including the main state arms export agency Rosoboronexport, were removed from U.S. lists of sanctioned entities.

Adopted on June 9, 2010,[31] Resolution 1929: are the following:[32]

- It added 15 Iranian firms affiliated with the Revolutionary Guard firms, and 22 other Iranian entities, to the list of U.N.-sanctioned entities. Some of the IRGC firms are alternate names for the *Khatem ol-Anbiya* (Seal of the Prophet) engineering firm under IRGC control.
- It made mandatory a ban on travel for Iranian persons named in it and in which a non-binding travel restriction was instituted in previous resolutions.
- It gave countries the authorization to inspect any shipments—and to dispose of their cargo—if the shipments are suspected to carry contraband items. However, inspections on the high seas are subject to concurrence by the country that owns that ship.
- It prohibited countries from allowing Iran to invest in uranium mining and related nuclear technologies, or in nuclear-capable ballistic missile technology.
- It banned sales to Iran of most categories of heavy arms and requests restraint in sales of light arms, but did not bar sales of missiles not on the "U.N. Registry of Conventional Arms" (meaning that the delivery of the S-300 system, discussed above, would not be legally banned).
- It required countries to insist that their companies refrain from doing business with Iran *if* such business could further Iran's WMD programs.
- It requested, but does not mandate, that countries prohibit Iranian banks to open in their countries, or for their banks to open in Iran, *if* doing so could contribute to Iran's WMD activities.
- It authorized the establishment of an eight person "panel of experts," with a mandate to assist the U.N. Sanctions Committee in implementing the Resolution and previous

Resolutions, and to suggest ways of more effective implementation. The Panel's coordinator is French; current membership includes the P-5 countries plus Japan, Germany, and Nigeria. The panel's reports are not officially published by the Sanctions Committee but are usually carried by various websites.

- An annex to the Resolution presented a modified offer of incentives for Iran to rejoin the international community.

2011-2012 Developments

President Obama and other senior officials noted that the intent of Resolution 1929 was to bring Iran back to negotiations. After its passage, EU foreign policy chief Catherine Ashton formally invited Iran to new talks. After several months of discussions, talks were held during December 6-7, 2010, with the P5+1, in Geneva, but by all accounts made little progress. The United States and Iran did not, as they did in the October 2009 talks, hold direct bilateral talks during the two days of meetings. Still, there was agreement to hold additional Iran-P5+1 talks in Istanbul which some thought might lead Iran to show more flexibility because of Turkey's willingness to take Iran's viewpoints into account.

The Istanbul (January 21-22, 2011) talks, by all accounts, made no progress because Iran demanded lifting of international sanctions as a precondition to substantive discussions. No date for new talks was announced. A P5+1 statement of March 9, 2011, affirmed the lack of progress at Istanbul, but also said "the door remains open."[33] Indications that talks might revive followed Iran-Russia talks during August 15-16, 2011, when Iran praised as a "basis to start negotiations" Russia's proposals for a stepwise exchange of the lifting of international sanctions for Iran's giving up some nuclear activities. A State Department spokeswoman, Victoria Nuland, confirmed that U.S. diplomats had worked with Russian counterparts to develop the proposal.

The prospect for new talks appeared to recede after the November 8, 2011, IAEA report cast doubt on Iran's assertions of its purely peaceful nuclear program. In late November 2011, in response to a move by the United States, Britain, and Canada to shut Iran out of the international banking system, a mob supported by the Basij militia ransacked the British Embassy in Tehran on November 29, 2011. This led to the closure of the Iranian and British embassies in London and Tehran, respectively, and caused the EU to impose on January 23, 2012, an embargo on purchases of Iranian oil, to be fully in effect by July 1, 2012. That decision complemented a December 2011 amendment to the FY2012 National Defense Authorization Act [P.L. 112-81] in pressuring Iran. (Canada closed its embassy in Iran in September 2012, citing security concerns.)

As sanctions started to harm Iran's economy severely, Iran—along with threats to the Strait of Hormuz—began to publicly propose a new round of nuclear talks. A letter of acceptance was sent by chief negotiator Jalilli to EU foreign policy chief Catherine Ashton on February 15, 2012. The United States and its partners announced on March 6, 2012, that they accept new talks.

April 13-14, 2012, Istanbul Talks

Talks were set for April 13-14, 2012, in Istanbul. Pressured by sanctions, Iran went into the talks suggesting it might accept P5+1 demands to halt 20% enrichment. The talks were held and, by all accounts, including a statement by EU foreign policy chief Ashton, did not focus on substantive details. However, Iran agreed to enter a negotiating process on its nuclear program—a pledge considered sufficient to announce a follow-up round in Iraq on May 23-24, 2012, to be held in Baghdad—a venue intended to accommodate Iran's earlier proposal. In preparation for the Baghdad talks, technical teams from the Iranian and P5+1 sides worked on substantive proposals.

May 23-24, 2012, Baghdad Talks

There was a sense of optimism going into the Baghdad talks, because Iranian leadership statements appeared to prepare the Iranian public for compromise by emphasizing that a prospective agreement preserves Iran's right to enrich uranium under the NPT. Tempering the optimism was the view among many experts that the Supreme Leader remains suspicions of U.S. and international intent and might, in the end, not accept an agreement. The following outlines what both sides, by numerous accounts, offered in Baghdad. The positions of both sides appear based on the principle of "reciprocity"—a term used by Secretary Clinton, EU foreign policy chief Ashton, and others—referring to a stepwise easing of sanctions in exchange for verifiable Iranian compliance.

The P5+1 reportedly proposed in Baghdad:

- That Iran halt enrichment to the 20% level and allow removal from Iran of the existing stockpile of 20% enriched uranium.
- That Iran eventually close the Fordow facility.
- That Iran accept a comprehensive verification regime to ensure that Iran fulfills any commitments made.
- That Iran clear up reputed past efforts to design a nuclear explosive device, including allowing inspections of Parchin and other facilities.

As "reciprocity" for Iran accepting such steps, the P5+1:

- Apparently would allow, at least in the interim, Iran to enrich uranium, although only to the 3.5%-5% level. This represents a dropping or de-emphasis of the demands of U.N. resolutions that Iran suspend all uranium enrichment.
- Offered Iran a guaranteed a supply of medical isotopes that it says it needs, and technical assistance to ensure the safety of its civilian nuclear facilities.
- Offered Iran spare parts for its civilian passenger aircraft.

Baghdad Outcome? The P5+1 did not offer to defer or lift entirely the EU oil embargo that goes into full effect July 1, leading to criticism in the Iranian media that the P5+1 offer was "unbalanced." This issue reportedly threatened the talks, but U.S. and other officials told journalists that the talks were substantive. According to EU foreign policy representative Ashton's statement at the conclusion of the talks, "it is clear that we both want to make progress, and that there is some common ground. However, significant differences remain. Nonetheless, we do agree on the need for further discussion to expand that common ground." According to the statement, Iran declared its readiness to address the 20% enrichment issue and presented a five part counter-proposal that focused on the nuclear issue but also raised the issue of U.S. and European involvement in Syria. There was enough progress to yield agreement that further Iran- P5+1 talks would be held June 18-19, 2012, in Moscow.

June 18-19 Moscow Talks and Follow-Up Discussions

Expectations were relatively low going into the Moscow talks, in that the P5+1 refused Iran's proposal to have technical talks immediately before the negotiations. However, Iran and EU chief negotiator Ashton reached agreement on an agenda for the talks, and Iran dropped threats to call the talks off. By all accounts, the talks were highly substantive, and Iran engaged specifically on the P5+1 demands, discussed above, to "stop, shut, and ship" (stop enriching to 20%, shut Fordow, and ship out the existing 20% enriched stockpile).

However, no breakthrough was achieved, and the parties were able to agree to hold lower level technical talks on July 3, 2012, in Istanbul. According to the Ashton statement concluding the meeting, the technical talks will further clarify the P5+1 proposal, increase P5+1 understanding of Iran's response, and study other issues raised in the talks. Subsequently, there would be contact between the deputies to Ashton and to Iranian chief negotiator Jalilli, followed by Ashton-Jalilli contact to discuss a possible further round of high level talks.

This roadmap has been followed to date, with technical talks on July 3 and then talks between Ms. Ashton's deputy, Helga Schmid, and Seyyed Jallili's deputy, Ali Baqeri, on July 24, 2012. Ms. Ashton and Seyyed Jallili spoke by telephone on August 2, 2012, agreeing to assess their positions during the Ramadan and European holiday periods and to subsequently speak by telephone at the end of August 2012. That latter conversation was delayed but Ashton and Jallili met in Istanbul on September 18, 2012, to assess the status of the technical discussions. However, no new P5+1 talks were announced as an outcome.

On the other hand, suggesting that sanctions are beginning to affect Iran's assessments, Iranian officials accompanying Ahmadinejad to the U.N. General Assembly meetings in late September 2012 reportedly floated to counterpart diplomats proposals to suspend 20% enrichment in exchange for an early lifting of the EU oil embargo. The P5+1 countries met on September 27, 2012, on the sidelines of the U.N. meetings, indicating a willingness to enter further negotiations but rejecting an early easing of sanctions. On October 9, 2012, Iran's Foreign Minister reformulated Iran's proposals as offering to limit Iran's enrichment (presumably to the 3.5—5% level) if the international community guaranteed a supply of 20% enriched uranium for its medical reactor, a supply the international community has said repeatedly they would guarantee.

The P5+1 countries are scheduled to meet on November 21, 2012, to coordinate positions in preparation for possible new talks in January 2013. Some reports suggest the P5+1 might discuss offering a broader, stepwise easing of sanctions if Iran agrees to the P5+1 proposals on 20% enrichment, enhanced inspections, and clearing up questions about possible past nuclear weapons research. This might reflect the view of experts who believe that Iran will not agree to the P5+1 demands unless and until the P5+1 offers to lift the EU embargo on purchases of Iranian oil, which took full effect on July 1, 2012. This has not been offered by the P5+1 to date.

Missiles and Chemical/Biological Weapons

Iran has an array of weapons that could pose problems for the United States and its allies in the Persian Gulf. In particular, Iran's missiles are considered to pose a threat to U.S. ships, forces, and allies in the Gulf region and beyond.

Table 6. Summary of Provisions of U.N. Resolutions on Iran Nuclear Program (1737, 1747, 1803, and 1929)

Requires Iran to suspend uranium enrichment.
Prohibits transfer to Iran of nuclear, missile, and dual use items, except for use in light-water reactors.
Prohibits Iran from exporting arms or WMD-useful technology.
Prohibits Iran from investing abroad in uranium mining, related nuclear technologies or nuclear capable ballistic missile technology.
Freezes the assets of over 80 named Iranian persons and entities, including Bank Sepah, and several corporate affiliates of the Revolutionary Guard.
Requires that countries ban the travel of over 40 named Iranians.
Mandates that countries not export major combat systems to Iran.
Calls for "vigilance" (a nonbinding call to cut off business) with respect to all Iranian banks, particularly Bank Melli and Bank Saderat.
Calls for vigilance (voluntary restraint) with respect to providing international lending to Iran and providing trade credits and other financing and financial interactions.
Calls on countries to inspect cargoes carried by Iran Air Cargo and Islamic Republic of Iran Shipping Lines—or by any ships in national or international waters—if there are indications they carry cargo banned for carriage to Iran. Searches in international waters would require concurrence of the country where the ship is registered.
A Sanctions Committee, composed of the fifteen members of the Security Council, monitors implementation of all Iran sanctions and collects and disseminates information on Iranian violations and other entities involved in banned activities. A seven member "panel of experts" is empowered to report on sanctions violations and make recommendations for improved enforcement.

Chemical and Biological Weapons

Official U.S. reports and testimony state that Iran maintains the capability to produce chemical warfare (CW) agents and "probably" has the capability to produce some biological warfare agents for offensive purposes, if it made the decision to do so.[34] This raises questions about Iran's compliance with its obligations under the Chemical Weapons Convention (CWC), which Iran signed on January 13, 1993, and ratified on June 8, 1997.

Ballistic and Cruise Missiles and Warheads

The Administration view is that Iran's growing inventory of ballistic missiles and its acquisition of indigenous production of anti-ship cruise missiles (ASCMs) provide capabilities for Iran to project power. Tehran views its conventionally armed missiles as an integral part of its strategy to deter—and if necessary retaliate against—forces in the region, including U.S. forces. On the other hand, U.S. defense officials say that U.S. forces are able to destroy these capabilities shortly after Iran uses them against U.S. assets in the Gulf region. A particular worry of U.S. commanders remains Iran's inventory of cruise missiles, which can reach U.S. ships in the Gulf quickly after launch.

The April 2012 Defense Department report, cited earlier, says that Iran is steadily expanding its missile and rocket inventories, and has "boosted the lethality and effectiveness of existing systems with accuracy improvements and new submunition payloads. This report appears to credit Iran's missile technology to a greater degree than past official reports. However, the report does not alter the longstanding U.S. estimate that Iran would likely not be able to fully develop a missile of intercontinental range until 2015. Secretary of Defense

Panetta said in January 2012 that Iran might be able to develop a nuclear-armed missile about a year or two after developing a nuclear explosive device.[35] The U.N. expert panel created by Resolution 1929 reported in May 2011 that might be getting ballistic missile technology from North Korea, in violation of U.S. sanctions against Iran. **Table 7** contains some details on Iran's missile programs.[36]

Table 7. Iran's Missile Arsenal

Shahab-3 ("Meteor")	800-mile range. The missile is operational, and Defense Department report of April 2012, indicates Tehran has improved its lethality and effectiveness, tempering previous assessments by experts that the missile is not completely reliable.
Shahab-3 "Variant"/ Sijil/ Ashoura	1,200-1,500-mile range. The April 2010 Defense Department report had the liquid fueled Shahab-3 "variant" as "possibly deployed," and the April 2102 report indicates the solid fuel version (Sijil or Ashoura) is increasing in range, lethality, and accuracy. These missiles potentially put large portions of the Near East and Southeastern Europe in range, including U.S. bases in Turkey. A U.N. experts panel reported in May 2011 that Iran tested the missile in October 2010 although the launch was "reported by a [U.N.] Member state," and not announced publicly. In concert with the beginning of 10-day "Great Prophet Six "military exercises, on June 28, 2011, Iran unveiled underground missile silos and undertook some missile tests.
BM-25	1,500-mile range. On April 27, 2006, Israel's military intelligence chief said that Iran had received a shipment of North Korean-supplied BM-25 missiles. Missile said to be capable of carrying nuclear warheads. The *Washington Times* appeared to corroborate this reporting in a July 6, 2006, story, which asserted that the North Korean-supplied missile is based on a Soviet-era "SS-N-6" missile. Press accounts in December 2010 indicate that Iran may have recei-ved components but not the entire BM-25 missile from North Korea.
ICBM	U.S. officials believe Iran might be capable of developing an intercontinental ballistic missile (3,000 mile range) by 2015, a time frame reiterated by the April 2012 DOD report.
Short Range Ballistic Missiles and Cruise Missiles	Iran is fielding increasingly capable, short range ballistic missiles, according to DOD 2012 report, such as ability to home in on and target ships while in flight. One version could be a short range ballistic missile named the Qiam, tested in August 2010. Iran has long worked on a 200 mile range "Fateh 110" missile (solid propellant), which it again tested in August 2012. Iran also possesses a few hundred short-range ballistic missiles, including the *Shahab-1* (Scud- b), the *Shahab-2* (Scud-C), and the *Tondar-69* (CSS-8). In January 2009, Iran claimed to have tested a new air-to-air missile. On March 7, 2010, Iran claimed it was now producing short- range cruise missiles that it claimed are highly accurate and can destroy heavy targets. Iran also is able to arm its patrol boats with Chinese-made C-802 cruise missiles. Iran also has Chinese-supplied Seerseekers and C-802's emplaced along Iran's coast.
Space Vehicle	In February 2008 Iran claimed to have launched a probe into space, suggesting its missile technology might be improving to the point where an Iranian ICBM is realistic. Following an August 2008 failure, in early February 2009, Iran successfully launched a small, low-earth satellite on a Safir-2 rocket (range about 155 miles). The Pentagon said the launch was "clearly a concern of ours" because "there are dual-use capabilities here which could be applied toward the development of long-range missiles." A larger space vehicle, Simorgh, was displayed in February 2010. Iran claimed to have launched a satellite into orbit on June 16, 2011.
Warheads	*Wall Street Journal* report of September 14, 2005, said that U.S. intelligence believes Iran is working to adapt the Shahab-3 to deliver a nuclear warhead. Subsequent press reports say that U.S. intelligence captured an Iranian computer in mid-2004 showing plans to construct a nuclear warhead for the Shahab.37 The IAEA is seeking additional information from Iran.

On November 12, 2011, an unexplained explosion at a ballistic missile base 25 miles from Tehran killed 17 IRGC missile force officers, including commander Hassan Tehrani Moghaddam. He reportedly was instrumental in developing Iran's missile force. Iran said the explosion was accidental, caused during a movement of munitions at the base. However, the death of Moqhaddam in the blast raised suspicions that it was caused by sabotage. The base was almost completely destroyed, according to commercial satellite photos posted on various websites.

FOREIGN POLICY AND SUPPORT FOR TERRORIST GROUPS

Iran's foreign policy is a product of the ideology of Iran's Islamic revolution, blended with long- standing national interests and the belief that the United States seeks to overturn Iran's Islamic revolution. Iran's leaders increasingly assert that Iran is a major regional power whose interests must be taken into account. Others interpret Iran's foreign policy objectives as beyond defensive—attempting to overturn the power structure in the Middle East that Iran believes favors the United States, Israel, and their "collaborators": Sunni Muslim regimes such as Egypt, Jordan, and the Gulf states. Iran couches that policy as support for an "oppressed" underclass in a region dominated by elites, while downplaying the underlying Iranian intent to empower fellow Shiites against the Sunni Muslims that dominate the region.

Some experts saw Iran as a potential strategic beneficiary of the uprisings that have toppled the leaders of Egypt, Tunisia, and Libya, and which have threatened the grip on power of the Sunni ruler of Bahrain. Some evidence of Iran as beneficiary could be provided by reports of a willingness of the new leaders of Egypt to end decades of diplomatic estrangement with Iran, and their decision to allow Iranian ships to transit the Suez Canal since the fall of President Mubarak. President Mohammad Morsi of Egypt attended the Non-Aligned Movement summit in Tehran during August 27-31, 2012. On the other hand, at the meeting, Morsi strongly denounced Syria's use of military action against its armed opposition, directly countering Iran's efforts at the meeting to increase international support for the Syrian government. Some assert that the holding of that meeting in Tehran, attended by nine heads of state and U.N. Secretary-General Ban Ki-Moon, as well as representatives of the nearly 120 countries in the NAM, illustrates that Iran is not isolated.

Others assert that, before Arab uprisings, Iran had benefitted from the installation of friendly regimes in Iraq and Afghanistan, and the continuing political strength of Hezbollah in Lebanon and Hamas in the Gaza Strip. On the other hand, Iran itself has become threatened by the unrest in its key Middle Eastern ally, Syria. In addition, countries in the region increasingly are helping the United States enforce strict sanctions against Iran.

A key Iranian policy objective is to position itself to strategically counter Israel. Ahmadinejad's various statements on Israel were discussed above, and Supreme Leader Khamene'i has repeatedly called Israel a "cancerous tumor." In December 2001, Rafsanjani, now considered a moderate, said that it would take only one Iranian nuclear bomb to destroy Israel, whereas a similar strike against Iran by Israel would have far less impact because Iran's population is large. Iran has hosted numerous conferences to which anti-peace process terrorist organizations were invited (for example: April 24, 2001, and June 2-3, 2002). The formal position of the Iranian Foreign Ministry, considered a bastion of moderates, is that Iran would not seek to block an Israeli-Palestinian settlement but that the process is too weighted toward Israel to yield a fair result. Iran supports Palestinian efforts to obtain U.N. recognition of statehood.

Support for International Terrorism

Iran's foreign policy often involves support of groups that are considered terrorist by the United States. The State Department report on international terrorism for 2011, released July 31, stated that Iran "remained an active state sponsor of terrorism in 2011," but did not repeat previous years' characterizations that it is "the most active state sponsor of terrorism." The report again cites the IRGC Qods Force as the primary instrument by which the regime supports militant movements abroad and uses these movements to implement Iran's foreign policy. The current Defense Minister of Iran is Brigadier General Ahmad Vahidi, the former Qods Forces commander.

In 2011 and 2012, U.S. officials have emphasized what they see as a new dimension to the Iranian threat—the potential for Iran to try to commit acts of terrorism in the United States itself. This was discussed by DNI James Clapper in testimony before the Senate Intelligence Committee on January 31, 2012, and represents a change from the previous U.S. view that the risk of U.S. retaliation makes Iran's leaders highly unlikely to authorize attacks inside the United States. The altered assessment is based on an alleged Iranian plot, revealed on October 11, 2011 by the U.S. Justice Department, to assassinate the Saudi Ambassador to the United States.

Table 8. Major Past Acts of Iran or Iran-Related Terrorism

Date	Incident/Event	Likely/Claimed Perpetrator
April 18, 1983	Truck bombing of U.S. Embassy in Beirut, Lebanon. 63 dead, including 17 U.S. citizens.	Factions that eventually formed Lebanese Hezbollah claimed responsibility.
October 23, 1983	Truck bombing of U.S. Marine barracks in Beirut. 241 Marines killed.	Same as above
December 12, 1983	Bombings of U.S. and French embassies in Kuwait City. 5 fatalities.	Da'wa Party of Iraq—Iran- supported Iraqi Shiite militant group. 17 Da'wa activists charged and imprisoned in Kuwait
September 20, 1984	Truck bombing of U.S. embassy annex in Beirut. 23 killed.	Factions that eventually formed Lebanese Hezbollah
May 25, 1985	Bombing of Amir of Kuwait's motorcade	Da'wa Party of Iraq
June 14, 1985	Hijacking of TWA Flight 847. One fatality, Navy diver Robert Stetham	Lebanese Hezbollah
April 5, 1988	Hijacking of Kuwait Air passenger plane. Two killed.	Lebanese Hezbollah, seeking release of 17 Da'wa prisoners in Kuwait.
March 17, 1992	Bombing of Israeli Embassy in Buenos Aires. 29 killed.	Lebanese Hezbollah, assisted by Iranian intelligence/ diplomats.
July 18, 1994	Bombing of Argentine-Jewish Mutual Association (AMIA) building in Buenos Aires.	Same as above
June 25, 1996	Bombing of Khobar Towers housing complex near Dhahran, Saudi Arabia. 19 U.S. Air Force personnel killed.	Saudi Hezbollah, supported by Iran. Some assessments point to possible involve-ment of Al Qaeda as well.

Source: CRS. Prepared with the assistance of Hussein Hassan, Knowledge Services Group.

A further trend in 2012 has been apparent Iranian-sponsored attempts to attack Israeli diplomats and citizens, perhaps in retaliation for what Iran alleges has been Israeli assassinations of Iranian nuclear scientists. India reportedly has concluded the Qods Force was responsible for wounding the wife of an Israeli diplomat in an attack in Delhi in February 2012. Israel says Iran, working through its Lebanon ally Hezbollah, was responsible for a July 19, 2012, terrorist bombing in Bulgaria that killed five Israeli tourists. Other alleged Iranian plots against Israeli and other targets have been reported in 2012 in Thailand, Georgia, Azerbaijan, Cyprus, and Kenya.

Earlier, Iranian terrorism also took the form of assassinating dissidents abroad. In the late 1980s and early 1990s, Iran allegedly was responsible for the assassination of several Iranian dissidents based in Europe, including Iranian Kurdish dissident leader Abdol Rahman Qasemlu, several other Kurdish leaders (including those killed at the Mykonos café in Berlin in September 1992), the brother of PMOI leader Masud Rajavi, and several figures close to the late Shah of Iran. In May 2010, France allowed the return to Iran of Vakili Rad, who had been convicted in the 1991 stabbing of the Shah's last prime minister, Shahpour Bakhtiar. Iran has not been accused of dissident assassinations abroad in well over a decade.

Foreign Policy: Relations with the Persian Gulf States

The Persian Gulf monarchy states (Gulf Cooperation Council, GCC: Saudi Arabia, Kuwait, Bahrain, Qatar, Oman, and the United Arab Emirates) are concerned about Iranian strategic influence in the Gulf. Several Gulf rulers have asserted, mostly privately, that the United States should move decisively to end Iran's nuclear program, and they are cooperating with U.S. containment and missile defense strategies and with many aspects of U.S. and multilateral sanctions against Iran, as discussed later in this paper. In particular, those Gulf states with spare oil production capacity are offering more oil to customers that are reducing purchases of Iranian oil. However, the Gulf states have not openly supported U.S. conflict with Iran, fearing doing so might cause Iran to retaliate against Gulf state targets, and they maintain relatively normal trade and other interactions with Iran. On May 14, 2012, the GCC states largely rejected a Saudi plan for greater political and economic integration of the GCC states—a plan proposed in part to present a united front against Iran – as likely to forfeit some of their individual sovereignty. The Gulf states also appear to be working toward the ouster of Syrian President Bashar al-Assad in part to weaken Iran strategically in the region.

- *Saudi Arabia.* Iran and Saudi Arabia represent opposite poles of influence and interests in the region and Saudi leaders have threatened to try to acquire a nuclear weapon if Iran acquires one. Saudi Arabia sees itself as leader of the Sunni Muslim world and views Shiite Muslims, including those in eastern Saudi Arabia, as heretical. Saudi alarm over Iranian influence in the Gulf was a major factor in the military intervention by Saudi Arabia (joined by UAE) on behalf of the Bahrain government in March 2011. The Saudis repeatedly criticize Iran for past actions, including inspiring violent demonstrations at some of the annual Hajj pilgrimages in Mecca in the 1980s and 1990s—and which caused a break in relations from 1987-1991), and for supporting a pro-Iranian movement in the Kingdom, Saudi Hezbollah, that the Saudis hold responsible for the June 25, 1996, Khobar Towers bombing.[38] There have been some Shiite demonstrations in Saudi Arabia since the Arab uprisings began in early 2011 but there is no hard evidence of any Iranian involvement in that unrest. At the same time, Saudi-Iran ties progressed to high-level contacts during the presidency of the relatively moderate Mohammad Khatemi, including visits by him to the Kingdom in 1999 and 2002, and Ahmadinejad has visited the Kingdom on several occasions. Iran's intelligence minister, Heydar Moslehi, visited in December 2011 to reduce tensions.

- *United Arab Emirates* (UAE) concerns about Iran have not recovered from the April 1992 Iranian expulsion of UAE security forces from the Persian Gulf island of Abu Musa, which it and the UAE shared under a 1971 bilateral agreement. (In 1971, Iran, then ruled by the U.S.-backed Shah, seized two other islands, Greater and Lesser Tunb, from the emirate of Ras al-Khaymah, as well as part of Abu Musa from the emirate of Sharjah.) The UAE seeks to refer the dispute to the International Court of Justice (ICJ), but Iran insists on resolving the issue bilaterally. (ICJ referral requires concurrence from both parties to a dispute.) The UAE formally protested Iran's setting up of a maritime and ship registration office on Abu Musa in July 2008. The issue reignited on April 11, 2012, when Ahmadinejad visited Abu Musa, causing the UAE to submit to Iran and to the United Nations a formal letter of protest. On May 2, 2012, when IRGC Commander Mohammad Ali Jaafari, accompanied by several Majles deputies, visited the island and discussed developing a tourism industry there. UAE officials say the visits undid about one year of quiet diplomacy between the two countries on the issue, which included the naming of negotiators by both sides. In November 2012, the IRGC Navy established a new base to reinforce its authority over the three disputed islands. The United States supports UAE proposals but takes no formal position on sovereignty.

  Earlier, seeking to avoid antagonizing Iran, in May 2007 the UAE received Ahmadinejad (the highest-level Iranian visit since the 1979 revolution) and allowed him to lead an anti-U.S. rally of several hundred Iranian-origin residents of Dubai at a stadium there. This large Iranian-origin resident community (about 300,000) in Dubai may explain why Dubai takes a generally softer line on Iran than does the federation capital, Abu Dhabi.

  In cooperating with U.S. and international sanctions, the UAE has increased scrutiny of exports to Iran and restricted Iranian banking activities in the UAE since the passage of Resolution 1929 in June 2010, and it has frozen the assets of Iranians subject to asset freezes under the U.N. resolutions. In February 2012, a major bank in Dubai (Noor Islamic Bank) ended transactions with Iran; Iran was using the bank to process hard currency oil payments. UAE officials say their cooperation with sanctions enforcement has reduced trade with Iran by about 30% since 2010 (when the level was about $12 billion). The UAE is also close to completing a new oil pipeline that will allow its oil exports to bypass the Strait of Hormuz and lessen the vulnerability of UAE oil exports.

- *Qatar* is wary that Iran might eventually seek to encroach on its large North Field (natural gas). It shares that field with Iran (called South Pars on Iran's side) and Qatar earns large revenues from natural gas exports from it. Qatar's fears have been heightened by occasional Iranian statements, such as one in April 2004, when Iran's deputy oil minister said that Qatar is probably producing more gas than "her right share" from the field and that Iran "will not allow" its wealth to be used by others. Possibly to try to ease such implied threats, Qatar invited Ahmadinejad to the December 2007 GCC summit there.

- *Bahrain* is about 60% Shiite-inhabited, many of whom are of Persian origin, but its government is dominated by the Sunni Muslim Al Khalifa family. In 1981 and again in 1996, Bahrain publicly accused Iran of supporting Bahraini Shiite dissidents in efforts to overthrow the ruling Al Khalifa family. These concerns underlie the government response to the 2011-2012 uprising against the Al Khalifa regime by mostly Shiite demonstrators. In November 2011, an investigatory commission (Bahrain Independent Commission of Inquiry) concluded there is no evidence to indicate Iran instigated the protests, although U.S. officials say there is evidence Iran is working with Shiite hardline groups to block a political settlement there. Tensions have flared several times since July 2007 over Iranian attempts to question the legitimacy of a 1970 U.N.-run referendum in which Bahrainis opted for independence from Iran.
- *Oman*. Of the GCC states, the Sultanate of Oman is closest politically to Iran and has tended not to directly criticize Iranian policies. During the Shah's rule, Iran sent troops to help the Sultan suppress rebellion in the Dhofar region. Sultan Qaboos made a state visit to Iran in August 2009, coinciding with the second inauguration of Ahmadinejad that coincided with substantial Iranian unrest inside Iran over his reelection. Oman played a brokering role in obtaining the release from Iran of U.S. hiker Sara Shourd in September 2010, and her companions in September 2011. Some press reports say Omani officials routinely turn a blind eye to or even cooperate in the smuggling of western goods to Iran.
- *Kuwait* had pursued ties to Iran as a counterweight to Saddam Hussein, who invaded Kuwait in August 1990. Since Saddam's overthrow in 2003, Kuwait has become more distant from and critical of Iran and, in May 2010, Kuwait confirmed that it had arrested some Kuwaiti civil servants and stateless residents for allegedly working on behalf of the Qods Force in an alleged plot to blow up Kuwaiti energy facilities.[39] In March 2011, a Kuwait court sentenced two Iranians and a Kuwaiti to death in the alleged plot. In a related development that month, Kuwait expelled three Iranian diplomats, and Iran expelled three Kuwaiti diplomats in response. Iranian official visits to Kuwait in May 2011 defused the issue, to some extent, but the suspicions flared again in November 2011 when Iran arrested two persons its says are Kuwaiti spies. About 25% of Kuwaitis are Shiite Muslims, and Iran supported Shiite radical groups in Kuwait in the 1980s as a means to try to pressure Kuwait not to support the Iraqi war effort in the Iran-Iraq War (1980-1988). Some of those acts are listed in the table above.

Yemen

Yemeni leaders have long claimed that Iran was trying to destabilize Yemen—making such claims perhaps in order to secure more financial and military aid from the Arab Gulf states. More recently there have been independent indications of a growing Iranian role inside Yemen. In the north, an unnamed U.S. official reportedly said that Iranian smugglers backed by the Quds Force are using small boats to ship in AK-47s, rocket-propelled grenades and other arms to replace older weapons used by the Houthi rebels.[40] On July 19, 2012, Yemen's President accused Iran of spying on Yemen and threatened unspecified retaliation if Iran continued to do so. The overlay of the conflict in Yemen is an uprising against longtime President Ali Abdullah Saleh that began in 2011 and led to his departure from Yemen in January 2012.

Iranian Policy in Iraq[41]

The U.S. military ousting of Saddam Hussein in 2003 benefitted Iran strategically by removing its long time antagonist and bringing to power a government led by Shiite Islamists with long- standing ties to Iran. The Iraqi government first formed in May 2006 is still led by Prime Minister Nuri al-Maliki, the head of a Shiite Islamist party. He has made numerous visits to Iran during his tenure and is strongly backed by Iran in his political disputes with Iraq's Sunni-dominated factions. He and allied Iraqi leaders are sympathetic to some of Iran's regional goals, such as keeping in power Iran's ally President Bashar al-Assad of Syria, and of eliminating Iranian opposition groups from Iran's borders. Iraq reportedly has allowed Iran to overfly Iraqi airspace with cargo flights to supply the Syrian military in its battle against armed dissidents,[42] although U.S. pressure on Iraq in September 2012 reportedly has led Baghdad to cease that permission. Still, some Sunni opponents of Maliki report that Iraq is allowing ground convoys of arms to proceed from Iran to Syria. Iraq also reportedly continues to conduct a full spectrum of trade with Iran, sometimes allegedly running afoul of U.S. and international sanctions against Iran. On July 31, 2012, the United States sanctioned the Elaf Islamic Bank of Iraq for conducting sanctionable banking transactions with Iran.

Aside from Iran's ties to Maliki and governing institutions, Iran exercises influence in Iraq through Shiite factions, particularly that of Shiite cleric Moqtada Al Sadr. Iran continues to support Sadrist and other pro-Iranian Shiite militias in Iraq—such as the Promised Day Brigade, As'aib Ahl Al Haq (League of the Righteous) and Kata'ib Hezbollah (Hezbollah Brigades). However, these and other Shiite militias are reported to be less active since the U.S. departure in December 2011—a departure that removed their rationale for remaining armed. Iran reportedly used its influence with Sadr to dissuade him from directing his 40 followers in the Iraqi parliament to side with factions seeking a vote of no-confidence against Maliki.

Another possible sign of Iranian intent to influence Iraq came in November 2011 when supporters of an Iraq-born senior cleric, longtime resident in Iran, Ayatollah Mahmoud Shahrudi, began opening representatives offices in the holy Iraqi city of Najaf. This is viewed as an Iranian effort to position Shahrudi as a successor to the frail senior Iraqi cleric, Grand Ayatollah Ali al-Sistani. Iraqi Shiites reportedly are shunning Shahrudi, suggesting they do not want additional Iranian religious influence in Najaf. Maliki met with Shahrudi during his April 2012 visit to Iran.

Supporting Palestinian Militant Groups

Iran's support for Palestinian militant groups has long concerned U.S. administrations, as part of an apparent effort by Tehran to obstruct an Israeli-Palestinian peace, which Iran believes would strengthen the United States and Israel. The State Department report on terrorism for 2011 repeated previous year's reports assertions that Iran provides funding, weapons, and training to Palestinian Islamic Jihad (PIJ), the Al Aqsa Martyr's Brigades, and the Popular Front for the Liberation of Palestine-General Command (PFLP-GC). All are named as foreign terrorist organizations (FTO) by the State Department for their use of violence against Israel.

Iran and Hamas

Since mid-2011, the Iran-Hamas relationship, once considered a key to Iran's influence in stoking Israeli-Palestinian conflict, appears to have deteriorated over their differing views of the unrest in Syria. Hamas opposes the military-led crackdown against unrest by Syrian President Bashar Al Assad, possibly out of sectarian sympathy with the mostly Sunni protesters in Syria. Because it took this position, Hamas' Syria-based leaders left that country in late 2011. This position is at odds with Iranian policy and, perhaps as a punishment, Iran reportedly suspended payments to Hamas as of July 2011. In an attempt to signal that the two continue to have good relations, the prime minister of the Hamas authority in Gaza, Ismail Haniya, visited Tehran on February 10, 2012. However, on March 6, 2012, Hamas leaders stated they would not necessarily retaliate against Israel, on Iran's behalf, if Israel undertook unilateral military action against Iranian nuclear facilities. The Iran-Hamas rift appears to affirm the basic underlying vulnerability of the relationship in the form of their sectarian difference.[43] For well over a decade, the State Department report on terrorism, including the report for 2011, has said that Hamas receives funding, weapons, and training from Iran. However, the 2011 report contains information as of the end of 2011 and may not fully reflect the apparent rift.

When the Iran-Hamas relationship was strong, Iran's regional policy was strengthened by Hamas successes, such as its victory in the January 25, 2006, Palestinian legislative elections, and even more so by Hamas's June 2007 armed takeover of the Gaza Strip. Iran provided material support to Hamas during the December 27, 2008-January 17, 2009, Israel-Hamas War in Gaza. Then Joint Chiefs Chairman Admiral Mike Mullen said on January 27, 2009, that the United States boarded but did not seize a ship carrying light arms to Hamas from Iran; the ship (the *Monchegorsk*) later went to Cyprus. On February 1, 2009, one of Hamas' main leaders, Khaled Meshal, publicly praised Iran for helping Hamas achieve "victory" over Israel in the conflict.[44] Iran joined in regional criticism of Israel for its May 31, 2010, armed inspection of a Turkish ship, carrying humanitarian goods, that attempted to evade Israel's naval blockade of Gaza. In March 2011, Israel intercepted a ship, the *Victoria*, off its coast, and seized a "large quantity" of mortars and C- 704 cruise missiles that Israel said were bound for Hamas in Gaza. These shipments appeared to violate Resolution 1747, which bans Iranian arms exports. Iran expressed support for Hamas as hostilities with Israel escalated in mid-November 2012, but, perhaps because it has more diplomatic regional weight than does Iran, Egypt has been a more vocal proponent of Hamas' position than has Iran.

Lebanese Hezbollah[45]

Lebanese Hezbollah is Iran's chief protégé movement in the region. That relationship began when Lebanese Shiite clerics of the pro-Iranian Lebanese Da'wa Party began to organize in 1982 into what later was unveiled in 1985 as Hezbollah. Iran's political, financial, and military aid to Hezbollah has helped it become a major force in Lebanon's politics; it remains politically close to Iran but is no longer seen as Tehran's "proxy." Acts of terrorism by the group and its antecedents are listed in the table above, but Hezbollah has largely forsaken acts of international terrorism in recent years, focusing instead on its role in Lebanon.

Hezbollah's attacks on Israeli forces in southern Lebanon contributed to an Israeli withdrawal in May 2000, but Hezbollah maintained military forces along the border. Although Iran likely did not instigate Lebanese Hezbollah to provoke the July-August 2006 war with Israel, Iran has long been its major arms supplier. Hezbollah fired Iranian-supplied rockets on Israel's northern towns during the fighting, including at the Israeli city of Haifa (30 miles from the border), and, more intensively, at cities within 20 miles of the Lebanese border.[46] During that conflict, on July 14, 2006, Hezbollah hit an Israeli warship with a C-802 sea-skimming missile probably provided by Iran. Iran's 50 Qods Force personnel in Lebanon reportedly advised Hezbollah during the conflict, although that number might have increased during the conflict to help Hezbollah operate the Iran-supplied weaponry. Even though Hezbollah reduced its overt military presence in southern Lebanon in accordance with the conflict-related U.N. Security Council Resolution 1701 (July 31, 2006), Hezbollah was perceived as a victor in the war for holding out against Israel.

Ahmadinejad advertised Iran's continued strong commitment to Hezbollah during his October 14-15, 2010, visit to Lebanon, the first by a president of the Islamic Republic, which included his visiting villages near the border with Israel. Iran was perceived as a political beneficiary of Hezbollah's decision in January 2011 to withdraw from the Lebanese cabinet, which led to the fall of the Hariri government and the formation of a government by Hezbollah-selectee Najib Makati, a Sunni Muslim. (Under a long-standing agreed political formula in Lebanon, the prime minister is a Sunni Muslim.)

However, there has been more vocal criticism of Hezbollah within and outside Lebanon because it continues to support its other key patron, Syrian President Bashar Al Assad, despite his violent crackdown against protesters in Syria. Because of the perceived vulnerability of Assad, and Hezbollah's linkage to his regime, Iran is reported to be trying to broaden its relationships in Lebanon, particularly with the Christian community. In May 2012, Iran's first vice president visited Lebanon with a large delegation and numerous proposals to fund development projects in areas inhabited by all of Lebanon's different sects and confessions. One controversial project is to provide $40 million for a dam to provide electricity to parts of Lebanon's Christian heartland. Lebanon's cabinet is considering the offer.[47]

Recent Arming and Funding

Since the 2006 conflict, Iran has resupplied Hezbollah with at least 25,000 new rockets, and[48] press reports in early 2010 said Hezbollah maintains a wide network of arms and missile caches around Lebanon. Among the post-war deliveries were 500 Iranian-made "Zelzal" (Earthquake) missiles with a range of 186 miles, enough to reach Tel Aviv from south Lebanon. In November 2009, Israel intercepted a ship that it asserted was carrying 500 tons of arms purportedly for Hezbollah. Iran also made at least $150 million available for Hezbollah to distribute to Lebanese citizens (mostly Shiite supporters of Hezbollah) whose homes were damaged in the Israeli military campaign.[49] Many experts believe Hezbollah might fire those rockets at Israel if Israel were to strike Iran's nuclear facilities.

As far as funding, the State Department terrorism report for 2008, released on April 30, 2009, specified Iranian aid to Hezbollah as exceeding $200 million in 2008, and said that Iran trained over 3,000 Hezbollah fighters in Iran during that year. The report for 2009 used similar figures for Iranian aid and training for Hezbollah but over an unspecified time frame. The State Department report for 2011 repeated the 2010 report's assertion that Iran "has provided hundreds of millions of dollars in support of Hezbollah and has trained thousands of Hezbollah fighters at camps in Iran." Some experts believe Iranian support to Hezbollah declined in 2011 as international sanctions have taken a toll on Iran's economy and supply of available hard currency.

Syria

Syria is Iran's closest Arab ally and Iran fears a strategic setback if the Sunni-led rebellion in Syria succeeds. Iran's relationship with Syria is key to Iran's efforts to support Hezbollah because Syria is the transit point for the Iranian weapons shipments to Hezbollah. Both Iran and Syria have used Hezbollah as leverage against Israel to achieve their regional and territorial aims.

To try to prevent Assad's downfall, Iran has materially supported the Syrian regime. U.S. officials told journalists in May 2011 that Iran was sending Qods Force advisers to help Syria crush the unrest. The April 2012 Defense Department report on Iran's military power, released in July 2012, said that "Iran probably has provided military trainers to advise Syrian security forces." On May 17, 2012, press reports quoted a study by the "panel of experts" that is monitoring Iranian compliance with the U.N. resolutions (see above) as saying that Iran has shipped weaponry to Syria (and Afghanistan).

The first tentative confirmation of Iran's presence in the Syria conflict came in January 2012, when the armed opposition in Syria captured several men who it said were members of the Qods Force, and showed them and their identification cards on a video. In late May 2012, senior Qods Force officer General Esmail Ghaani publicly confirmed in Iranian media that there are Qods Force personnel in Syria, saying they have helped stop attacks on civilians carried out by Syrian armed rebels. IRGC Commander-in-Chief Jafari affirmed that assertion on September 16, 2012. On August 28, 2012, an IRGC commander said Iran is "involved in fighting every aspect of a war, a military one in Syria" and a cultural one as well.

In March 2011, shortly after the uprising in Syria broke out, Turkey intercepted and removed weapons from an Iranian cargo plane bound for Syria. Turkey impounded another such shipment consisting of several truckloads in late April 2011. On April 14, 2011 and on several occasions since, U.S. officials have said that Iran is providing Syria with equipment to suppress crowds and to monitor and block protester use of the Internet.[50]

On the other hand, there are signs of strains in Iran over the aid to Syria. Some reports indicate that the architect of the support effort, Qods Force commander Qasem Soleimani, has come under criticism for funneling substantial Iranian state resources to the Syria effort—criticism stoked by Iran's economic problems laid bare in its early October currency collapse.[51] Other criticism stemmed from a threat in early October 2012 by Syrian rebels to execute 48 Iranians they captured in Syria and claim are Qods personnel. Perhaps to address these criticisms, about 275 Qods Force personnel reportedly were withdrawn from Syria in early October.

At the same time, perhaps trying to hedge its position should President Bashar Al Assad fall, Iran has tried to encourage President Assad to take steps to calm the unrest through reforms. Foreign Minister Salehi said on August 28, 2011, that Syria should recognize its people's "legitimate demands." In September 2011, Ahmadinejad called on Assad to cease using force against demonstrators and open dialogue on reforms, and he and other Iranian leaders have expressed similar themes since. In late January 2012, Iran said Syria should hold free elections, but that doing so would take time to organize. In early November 2012, Iran announced it would host representatives of different ethnic and political groups from Syria, including from the Syrian government, in a "national dialogue;" no representatives of the armed opposition would be invited and the meeting was largely dismissed as an Iranian effort to support the Assad regime.

Iran bases its justification for its intervention in Syria on a longstanding defense relationship with the Assad regime. On December 13, 2009, the Syrian and Iranian defense ministers signed a defense agreement to "face common enemies and challenges." In late June 2010, it was reported that Iran had sent Syria a sophisticated air defense radar system that Syria could potentially use to thwart Israeli air strikes.[52] In March 2011, Iranian officials commented that they might contribute to improving some Syrian port facilities or other installations. On some occasions, including the early 1990s, Iran purportedly has acted as an intermediary with North Korea to supply Syria with various forms of WMD and missile technology. Iran reportedly has helped Syria expand its chemical weapons arsenal.[53]

In response to the Iranian military presence in Syria, on May 4, 2011, the Treasury Department designated the Qods Force as an entity subject to a U.S. assets freeze for human rights abuses in Syria (under Executive Order 13572). On May 18, 2011, the Treasury Department designated Mohsen Chizari, a Qods Force officer, and Qods Force overall commander Qasem Soleimani under that order. Other Iranians were sanctioned in late June under that order. In late August 2011, the European Union sanctioned the Qods Force for assisting the Syrian crackdown.

The Caucuses and Central Asia

Iran's policy in the nearby Caucuses has thus far emphasized Iran's rights to Caspian Sea resources, particularly against Azerbaijan. That country's population, like Iran's, is mostly Shiite Muslim, but its leadership is highly secular. Azerbaijan is ethnically Turkic, and Iran fears that Azerbaijan nationalists might stoke separatism among Iran's large Azeri Turkic population, which demonstrated some unrest in 2006 and during the uprising in Iran in 2009. These factors could explain why Iran has generally tilted toward Armenia, which is Christian, and which is at odds with Azerbaijan over territory and control of ethnic Armenians. In May 2012, Azerbaijan refused entry to a senior aide to Khamene'i. Iran has often slowed or stopped Azerbaijani truck traffic that must transit Iran in order to reach a non-continguous part of Azerbaijan (Nakichevan), which is cut off from the rest of Azerbaijan by Armenia's occupation of Nagorno-Karabakh.

In July 2001, Iranian warships and combat aircraft threatened a British Petroleum (BP) ship on contract to Azerbaijan out of an area of the Caspian that Iran considers its own. The

United States called that action provocative, and has since been engaged in border security and defense cooperation with Azerbaijan. The United States successfully backed construction of the Baku- Tblisi-Ceyhan oil pipeline, intended in part to provide alternatives to Iranian oil.

Israel also is apparently looking to Azerbaijan to counter Iran, announcing in February 2012 a major sale of defense equipment. In mid-March 2012, Azerbaijan arrested 22 persons it said were Iranian agents plotting attacks against Israeli and Western targets there.

Along with India and Pakistan, Iran has been given observer status at the Central Asian security grouping called the Shanghai Cooperation Organization (SCO—Russia, China, Kazakhstan, Kyrgyzstan, Uzbekistan, and Tajikistan). In April 2008, Iran applied for full membership in the organization, but, not wanting to antagonize the United States, in June 2010 the SCO denied Iran's bid by barring admission to countries under U.N. Security Council sanctions. Ahmadinejad attended the latest SCO meeting in Beijing on June 6-7, 2012, which he primarily used to press his case with China and Russia for relief from international sanctions over its nuclear program.

South Asia: Afghanistan, Pakistan, and India[54]

Iran looks to its eastern neighbors in South Asia as allies and potential allies to help parry U.S. and European pressure on Iran's economy and its leaders.

Afghanistan

In Afghanistan, as in Iraq, Iran is viewed by U.S. officials as pursuing a multi-track strategy— attempting to help develop Afghanistan and enhance its influence there, while also building leverage against the United States by arming anti-U.S. militant groups. Iran's main goal appears to be to restore some of its traditional sway in eastern, central, and northern Afghanistan, where Persian-speaking Afghans predominate. Many Afghans, even those of Pashtun ethnicity, speak Dari, a dialect of Persian language. Reports in mid-2012 suggest that Iranian traders are conducting financial or currency transactions in Afghanistan in an effort to acquire U.S. dollars and blunt the effects of international sanctions against Iran.[55]

Iran has sought some influence by supporting the government of President Hamid Karzai, who is a Sunni Muslim and a Pashtun. Karzai has said publicly, most recently in December 2011 (during a visit by Defense Secretary Panetta) that he does not want proxy competition between the United States and Iran in Afghanistan. Karzai regularly meets with Ahmadinejad bilaterally and in the context of several regional summit series that include Pakistan and Central Asian states. The latest such summit, between Iran, Afghanistan, and Pakistan, was held during February 17-18, 2012. Karzai admitted on October 26, 2010, that press reports were true that Iran has given Afghanistan direct cash payments (about $2 million per year) to support its budget. In addition, the two countries are said to be cooperating effectively against narcotics trafficking from Afghanistan into Iran; Iranian border forces take consistent heavy losses in operations to try to prevent this trafficking.

While dealing with Karzai, Iran also is positioning itself—in ways at odds with Afghan government interests—to threaten U.S. forces. Reflecting concern about the U.S. military presence in Afghanistan, Iran reportedly tried to derail the U.S.-Afghanistan Strategic

Partnership Agreement that was signed on May 1, 2012, and which provides for the possibility that the United States will keep forces in Afghanistan beyond the planned transition to Afghan lead in 2014. The agreement prohibits the United States from using Afghanistan as a base from which to launch military action against other countries. Still, U.S. officials say Iran tried to sway Afghan parliamentarians against the pact with bribes, but the Afghan legislators accepted Iran's payments but voted in favor of the pact anyway. Others say Iran is capable of mobilizing thousands of Afghans, particularly from the west of Afghanistan, to support Iranian interests there.

Iran is particularly concerned about the U.S. use of the air base at Shindand, in Herat Province, which Iran believes could be used for surveillance of or strikes on Iran. The U.S. drone that crashed in Iran in December 2011 was believed to be based in Shindand. The State Department terrorism report for 2011 continued to accuse the Qods Force of supplying various munitions, including 107 mm rockets, to select Taliban and other militants in Afghanistan, and of training Taliban fighters in small unit tactics, small arms use, explosives, and indirect weapons fire. The report again asserts that Iran has supplied militants in Qandahar, which is a Pashtun-inhabited province in southern Afghanistan and demonstrates that Iran is not only assisting militants near its borders. In February 2011, British forces captured a shipment of 48 Iranian-made rockets in Afghanistan's western province of Nimruz, allegedly bound for Taliban militants. On the other hand, U.S. commanders have consistently maintained that the Iranian assistance to Afghan militants is not decisive on the battlefield.

Iran's materiel support for Taliban elements may also give it leverage in any Taliban-government political settlement in Afghanistan; Iran reportedly invited some Taliban members to an "Islamic Awakening" conference in Tehran in mid-September 2011. The invitation was reportedly part of an attempt to broker a meeting between the Taliban representatives and Afghan government representatives attending the conference—the chief Afghan representative was the then head of the Afghanistan High Peace Council overseeing the reconciliation process, former President Burhanuddin Rabbani who was assassinated in Kabul after returning from the conference. In July 2012, Iran reportedly allowed the Taliban to open an office in Zahedan, in eastern Iran—possibly to better coordinate policy with the Taliban or possibly to facilitate and Iranian role in political reconciliation in Afghanistan.[56]

In August 2010, the Treasury Department sanctioned two Iranian Qods Force officers allegedly involved in supplying funds and materiel to Afghan militants. They were sanctioned under Executive Order 13224 for supporting international terrorism.

Because Iran has multifaceted influence in Afghanistan, some U.S. officials reportedly are arguing that the United States should develop a bilateral dialogue on Afghanistan, to be conducted by their respective ambassadors in Kabul. Iran may have signaled a willingness for such engagement when it attended the October 18, 2010, meeting in Rome of the 44-nation "International Contact Group" on Afghanistan. The United States did not object to the Iranian attendance at the meeting, which included a briefing by General David Petraeus (then top U.S./NATO commander in Afghanistan). Iran also attended the a Contact Group meeting on March 3, 2011, in Jeddah, Saudi Arabia (at the headquarters of the Organization of Islamic Conference). Iran did not attend the January 28, 2010, international meeting in Britain on Afghanistan, but it did attend a follow-up meeting in Kabul on July 20, 2010. Iran attended the regional meeting on Afghanistan in Istanbul on November 2, 2011, at which all regional countries pledged to support Afghan stability and sovereignty. It also attended the major international Bonn Conference on Afghanistan on December 5, 2011, and subsequent international conferences on that issue (with the obvious exception of the May 20-21, 2012, NATO summit in Chicago).

Pakistan

Iran's relations with Pakistan have been partly a function of events in Afghanistan, although relations have worsened somewhat in late 2009 as Iran has accused Pakistan of supporting Sunni Muslim rebels in Iran's Baluchistan region. These Sunni guerrillas have conducted a number of attacks on Iranian regime targets in 2009, as discussed above (*Jundullah*).

Iran engaged in substantial military cooperation with Pakistan in the early 1990s, and it was revealed in 2003 that the founder of Pakistan's nuclear weapons program, A. Q. Khan, sold Iran nuclear technology and designs.[57] However, Iran-Pakistan relations became strained in the 1990s when Pakistan was supporting the Taliban in Afghanistan, which committed alleged atrocities against Shiite Afghans (Hazara tribe), and which seized control of Persian-speaking areas of western and northern Afghanistan. Iran remains suspicious that Pakistan might want to again implant Taliban militants in power in Afghanistan, but Iran-Pakistan relations have improved since mid-2011 as Pakistan's relations with the United States have worsened. Iran and Pakistan now have a broad bilateral agenda that includes a potential major gas pipeline project. Iran has completed the line on its side of the border, and Pakistan is trying to accelerate work on its part of the line. However, Pakistan reportedly has had trouble financing the project on its side of the border, and U.S. officials say they consider it potential sanctionable under the Iran Sanctions Act.

India

India and Iran have sought to accommodate each others' interests and avoid mutual conflict. Their interests have tended to align on several issues, particularly Afghanistan, where both countries support the minority factions based in the north and west. India reportedly wants to expedite the development of Iran's Chabahar port, which would give India direct access to Afghanistan and Central Asia without relying on transit routes through Pakistan.

As international sanctions have increased in 2011-2012, India appears to be wrestling with a choice of preserving its ties to Iran—which has provided it with needed oil for its growing economy—or joining U.S. and international attempts to isolate Iran. In 2012, it has sided with the United States and the EU by cutting its purchases of Iranian oil, and has received an exemption from U.S. sanctions. India's participation in the trans-Pakistan gas pipeline discussed above appears unlikely, at least initially, not only for fear of potential U.S. sanctions but also because of India-Pakistan mistrust but also over pricing and security issues. India and Iran reportedly have discussed an alternative undersea pipeline that would bypass Pakistan. Perhaps to balance these actions in support of sanctions, Prime Minister Manmohan Singh attended the Non-Aligned Movement summit in Tehran in August 2012.

Of particular concern to some U.S. officials, particularly in the late 1990s, were India-Iran military-to-military relationships and projects. The relationship included visits to India by some Iranian naval personnel, although India said these exchanges involve junior personnel and focus mainly on promoting interpersonal relations and not on India's provision to Iran of military expertise. The military relationship between the countries has withered over at least the past five years. India and Iran, along with the United States, backed anti-Taliban factions in Afghanistan during 1996-2001.

Al Qaeda

Iran is not a natural ally of Al Qaeda, largely because Al Qaeda is an orthodox Sunni Muslim organization. However, some experts believe that hardliners in Iran still might want to use Al Qaeda activists as leverage against the United States and its allies, despite the May 1, 2011, death of Al Qaeda founder Osama bin Laden in a U.S. raid in Pakistan. Some allege that Iran is forging links to Al Qaeda affiliates in Africa, such as Boko Haram in Nigeria and Al Shabab in Somalia, for the purpose of extending its influence in Africa. However, it is unclear whether any Iranian ties to these groups have been approved at the highest levels of the Iranian leadership.

The 9/11 Commission report said several of the September 11 hijackers and other plotters, possibly with official help, might have transited Iran, but the report did not assert that the Iranian government cooperated with or knew about the plot. A U.S. district court filing in May 2011 in New York named Iranian officials and ministries as materially supporting the Al Qaeda in the September 11 attacks. The court found, on December 15, 2011, in favor of the plaintiffs and later ordered Iran, Al Qaeda, and the Taliban to pay $6 billion in damages to the relatives of the September 11 attacks. Earlier, on November 28, 2011, a U.S. district court issued a ruling linking Iran (and Sudan) to the August 1998 Al Qaeda bombings of the U.S. embassies in Kenya and Tanzania.

Even though bin Laden has been killed, Iran might see possibilities for tactical alliance with Al Qaeda. Three major Al Qaeda figures believed to still be based mostly in Iran include spokesman Sulayman Abu Ghaith, top operative Sayf Al Adl, and a bin Laden's son, Saad.[58] All three have been reported, at least on occasion, to have been allowed outside Iran to travel to Pakistan. (U.S. officials blamed Saad bin Laden, Adl, and Abu Ghaith for the May 12, 2003, bombings in Riyadh, Saudi Arabia, against four expatriate housing complexes, saying they have been able to contact associates outside Iran.)[59] The Department of Treasury, on January 16, 2009, designated four Al Qaeda operatives in Iran, including Saad bin Laden (and three lesser known figures) as terrorist entities under Executive Order 13224. On July 28, 2011, under that same order, the Treasury Department sanctioned six Al Qaeda members for allegedly moving funds to Al Qaeda in Pakistan and Afghanistan via their bases in Iran, and under a specific agreement between Al Qaeda and Iran. Another bin Laden ally, Abu Musab al-Zarqawi, killed by U.S. forces in Iraq on June 7, 2006, reportedly transited Iran into Iraq after the September 11, 2001 attacks and became an insurgent leader in Iraq.

Iran has, to some extent, confirmed the presence of Al Qaeda militants in Iran. It asserted on July 23, 2003, that it had "in custody" senior Al Qaeda figures. On July 16, 2005, Iran's intelligence minister said that 200 Al Qaeda members are in Iranian jails.[60] U.S. officials have said since January 2002 that Iran has not prosecuted or extradited any senior Al Qaeda operatives. In December 2009, Iran's foreign minister confirmed that a teenage daughter of Osama bin Laden had sought refuge in the Saudi embassy in Tehran—the first official confirmation that members of bin Laden's family have been in Iran. She left Iran in March 2010. As many as 20 other family members have been said to be living in a compound in Iran since the September 11, 2001, attacks, and accusing Iran of refusing to allow them to leave for Saudi Arabia or other places. Some family members have said the young bin Ladens have never been affiliated with Al Qaeda.

Latin America

A growing concern in Congress has been Iran's developing relations with countries and leaders in Latin America. Iran views some Latin American countries, particularly Cuba and Venezuela, as sharing its distrust of the United States and as willing to help Iran circumvent some international sanctions. Suggesting expanded Iranian interest in the Western Hemisphere, during 2006-2011, Iran opened six embassies in countries in the region (Colombia, Nicaragua, Chile, Ecuador, Uruguay, and Bolivia), and expanded embassies in Cuba, Argentina, Brazil, Mexico, and Venezuela. In January 2012, Ahmadinejad undertook a visit to Latin America, including Venezuela, Ecuador, Cuba, and Nicaragua. By all accounts, few concrete economic agreements were reached during that visit, which expands on past patterns in which agreements tend to be announced but not implemented. Ahmadinejad attended the U.N. Conference on Sustainable Development in Brazil on June 21, 2012, which was bounded by his travel to Bolivia and Venezuela.

On March 1, 2012, a House Foreign Affairs Committee Subcommittee on Terrorism, Nonproliferation and Trade reported out H.R. 3783, the "Countering Iran in the Western Hemisphere Act," requiring the Administration to develop a strategy to counter Iran's purportedly growing influence in Latin America. It was reported out by the full committee on March 7, 2012. Many outside experts, however, believe that most Latin American countries see little benefit to a major expansion of ties to Iran and that assessments of Iran's influence in the region may be overstated by some think-tank experts. No Latin American leader or head of government attended the NAM summit in Tehran in August 2012.

Venezuela and Cuba

Venezuela under President Hugo Chavez continues to be Iran's main champion in the region. Iranian leaders were undoubtedly relieved at Chavez's re-election on October 7, 2012; his opponent in the election, Henrique Capriles, was considered likely to align Venezuela closer to the United States and end Chavez's close ties to Iran. Still, there is no consensus on the degree of threat posed by Iran-Venezuela ties; in July 2012, President Obama stated that Iran-Venezuela ties do not constitute a strategic threat to the United States. The April 2012 Defense Department report on Iran does not address this issue at all, although the 2010 version of the report was the first U.S. government publication to say that Qods Force personnel are in Venezuela, where their presence had "increased" in recent years, according to that report.

President Chavez has visited Iran on several occasions, offering to engage in joint oil and gas projects, and Ahmadinejad has visited Venezuela on each of his six trips to Latin America as president, including the June 2012, trip. However, contrary to the assertions of some experts, a State Department official testified before the House Foreign Affairs Committee on June 24, 2011, that Iran's embassy in Caracas has only about 14 diplomats and is not particularly active in terms of open diplomatic activity, casting doubt on reports that Iran has a large, active presence in Venezuela. About 400 Iranian engineers have reportedly been sent to Venezuela to work on infrastructure projects there. It was reported in May 2011 that the two may have signed an agreement in October 2010 to develop a joint missile base in Venezuela, but Venezuela has denied these reports and the Obama Administration has said there is no evidence to support the missile base assertion. Venezuela reportedly has purchased some Iranian military equipment, such as rifles, as well as $23 million in military equipment upgrades and an explosives factory.[61]

Many accounts say that most of the agreements between Iran and Venezuela are agreements in principle that have not been implemented in reality. Among the arrangements implemented are the establishment of direct air links through an obscure air service dedicated to this route, although the route was suspended in 2010. A firm deal for Petroleos de Venezuela to supply Iran with gasoline was signed in September 2009, apparently in a joint effort to circumvent the reduction in worldwide sales of gasoline to Iran. In part because of this trade, the firm was sanctioned under the Iran Sanctions Act in May 2011.

Cuba

Iran's relations with Cuba are long-standing and Cuba has routinely been included in Ahmadinejad's several visits to Latin America. In the past, Cuba reportedly has helped Iran jam the broadcasts of Iranian dissidents based in Los Angeles and elsewhere in the United States. Still, Cuba's economy is widely considered too small to be able to materially reduce the effect of international sanctions against Iran.

Other Ties in the Region

Iran also has built ties to Ecuador, Bolivia, Nicaragua, and Brazil, although some press accounts may exaggerate the extent and strategic significance of these relations. Iranian Defense Minister Ahmad Vahidi visited Bolivia in May 2011, but President Evo Morales was then compelled to apologize to Argentina for inviting him because of Vahidi's alleged involvement in the 1994 Buenos Aires bombing listed in the table above. Vahidi was, at the time, the head of the Qods Force. Iran reportedly has $1 billion in joint ventures with Bolivia. These ventures reportedly were the subject of discussion during Ahmadinejad's June 2012 visit, discussed above. Trade with Ecuador expanded from $6 million annually to $168 million from 2007 to 2008. Iran's embassy in Managua, Nicaragua, is said by close observers to be small, contrasting with some reports on that issue, and Nicaragua has refused Iranian demands to repay $164 million in debt it owes Iran for past crude oil deliveries. Nicaragua reportedly was upset that Ahmadinejad's January 2012 visit did not result in an Iranian pledge to forgive that debt. Iran also failed to implement some promises to undertake joint ventures with Nicaragua, including a $350 million deep water port there. Still, President Daniel Ortega hosted Ahmadinejad during his visit there in January 2012.

Because of its large economy, Brazil, under previous President Luiz Inacio Lula da Silva, emerged as Iran's most significant supporter, particularly because of Brazil's engagement with Iran to forge the "Tehran Declaration" on nuclear issues in June 2010. However, the government of President Dilma Roussef, whose term began January 1, 2011, has been less supportive of Iran than was her predecessor. Ahmadinejad did not visit Brazil during his January 2012 visit to the region, furthering the impression that Iran's influence in South America is limited to mainly smaller nations. However, the mid-April 2012 statement that Ahmadinejad would "soon" visit Brazil appeared to be an effort by Tehran to show that its relations with Brazil are not in decline.

Africa

To reduce Iran's isolation, Ahmadinejad has reached tried to enlist the support of some African leaders. Some observers believe that Iran's outreach is focused on those African countries that might be able to export natural uranium for Iran's nuclear program to compensate for Iran's domestic deficiencies; such uranium producers include Zimbabwe, Senegal, Nigeria, and the Democratic Republic of Congo. In April 2010, Ahmadinejad visited Uganda and Zimbabwe, even though Zimbabwe's leader, Robert Mugabe, has himself been heavily criticized by the international community in recent years. Iran has also cultivated Senegal as an ally. Still, it is believed that African support for Iran is unlikely to outweigh Iran's growing estrangement from Europe and other regions. Ten heads of state or government attended the NAM summit in Tehran in August 2012, including Mugabe.

As an example of what the Administration called Iran's exports of lethal aid to foment violence in Africa, in October 2010, the Qods Force reportedly attempted to ship weapons to Gambia, via Nigeria, but the shipment was intercepted in Nigeria. Several Iranian entities, and a Nigerian shipping agent, were sanctioned by the United States in April 2012 for facilitating this incident. The Nigerian shipping agent allegedly helped Qods Force personnel enter Nigeria. U.N. panel of experts report on Iranian arms sales embargo violations, discussed above, have cited Iranian attempts to ship weapons to allies in the Middle East via Nigeria.

Some Members of Congress are concerned that Iran is supporting radical Islamist movements in Africa. Allegations of Iran's support for Al Qaeda affiliates in Africa are discussed above in the section on Al Qaeda. There have been allegations that Iran might have supplied Al Shabab in Somalia with anti-aircraft and anti-tank weaponry, although the few press reports about such activity suggests it is a minor component of Iranian policy, if at all.

Iran also appears to have an ongoing relationship with the government of Sudan. Relations reportedly were close in the early 1990s when Islamist leaders in Sudan were attempting to make Sudan into a hub for international Islamist movements. The relationship apparently cooled in the mid-1990s when international sanctions compelled Sudan to expel Osama bin Laden and to downplay its Islamist links abroad. Still, Iran is said to supply the Sudan government with weapons it is using on its various fronts, such as the one with South Sudan, and the Qods Force reportedly continues to train the Popular Defense Force militia. Some observers say Iranian pilots may be active in Sudan on behalf of the government there. President Omar Hassan Al Bashir attended the NAM summit meeting in Tehran in August 2012. On October 31, 2012, two Iranian warships docked in Port Sudan for joint exercises with Sudan, one week after a weapons factory in Khartoum was bombed, allegedly by Israel. the factory purportedly was a source of Iranian or other rockets and other weapons intended for Hamas. After the ship visit, some Sudanese politicians questioned the wisdom of Sudan's drawing closer to Iran.

U.S. Policy Approaches and Additional Options

The February 11, 1979, fall of the Shah of Iran, a key U.S. ally, opened a long and deep rift in U.S.-Iranian relations. Although U.S. concerns about Iran and its nuclear program are long- standing, a growing Israeli threat to use military action against Iran's nuclear facilities—with or without U.S. backing—has made U.S. policy toward Iran an urgent issue. Many of the policy options being implemented or under consideration are the same options that have faced the United States since 1979—and virtually no policy option has been taken "off the table."

Background on Relations Since the 1979 Revolution

The Carter Administration sought a degree of engagement with the Islamic regime during 1979, but it agreed to allow in the ex-Shah for medical treatment and engaged some moderate Iranian officials of the new regime who were viewed by Khomeini loyalists as insufficiently revolutionary. As a result, the U.S.-Iran estrangement began in earnest on November 4, 1979, when radical pro-Khomeini "students in the line of the Imam (Khomeini)" seized the U.S. Embassy in Tehran and held its diplomats hostage until minutes after President Reagan's inauguration on January 20, 1981. The United States broke relations with Iran on April 7, 1980 (two weeks prior to the failed U.S. military attempt to rescue the hostages during April 24-25, 1980), and the two countries had only limited official contact thereafter.[62]

The United States tilted toward Iraq in the 1980-1988 Iran-Iraq War, including U.S. diplomatic attempts to block conventional arms sales to Iran, providing battlefield intelligence to Iraq[63] and, during 1987-1988, direct skirmishes with Iranian naval elements in the course of U.S. efforts to protect international oil shipments in the Gulf from Iranian mines and other attacks. In one battle on April 18, 1988 ("Operation Praying Mantis"), Iran lost one-quarter of its larger naval ships in a one-day engagement with the U.S. Navy, including one frigate sunk and another badly damaged. Iran strongly disputed the U.S. assertion that the July 3, 1988, U.S. shoot down of Iran Air Flight 655 by the U.S.S. *Vincennes* over the Persian Gulf (bound for Dubai, UAE) was an accident.

After the Iran-Iraq War ended, President George H. W. Bush laid the groundwork for a rapprochement with Iran. In his January 1989 inaugural speech, saying that, in relations with Iran, "goodwill begets goodwill," implying better relations if Iran helped obtain the release of U.S. hostages held by Hezbollah in Lebanon. Iran reportedly did assist in obtaining their releases, which was completed in December 1991, but no thaw followed, possibly because Iran continued to back groups opposed to the U.S.-sponsored Middle East peace process.

Upon taking office in 1993, the Clinton Administration moved to further isolate Iran as part of a strategy of "dual containment" of Iran and Iraq. In 1995 and 1996, the Clinton Administration and Congress added sanctions on Iran (a ban on U.S. trade and investment with Iran and the Iran Sanctions Act that penalizes foreign investment in Iran's energy sector) in response to growing concerns about Iran's weapons of mass destruction and its efforts to subvert the Arab-Israeli peace process. The intent of these sanctions was to persuade U.S. allies to restrict trade with Iran. The Administration expressed skepticism of the EU's policy of "critical dialogue" with Iran—a dialogue consisting of EU meetings with Iran that included criticisms of Iran's human rights policies and its support for militant movements in the Middle East.

The election of Khatemi in May 1997 precipitated a U.S. shift toward engagement; the Clinton Administration offered Iran official dialogue, with no substantive preconditions. In

January 1998, Khatemi publicly agreed to "people-to-people" U.S.-Iran exchanges, but ruled out direct talks. In a June 1998 speech, then-Secretary of State Albright called for mutual confidence building measures that could lead to a "road map" for normalization. Encouraged by the reformist victory in Iran's March 2000 *Majles* elections, Secretary Albright, in a March 17, 2000, speech, acknowledged past U.S. meddling in Iran, announcing an easing of the U.S. trade ban with Iran, and promised to try to resolve outstanding claims disputes. In September 2000 U.N. "Millennium Summit" meetings in New York, Albright and President Clinton sent a positive signal to Iran by attending Khatemi's speeches.

George W. Bush Administration Policy

With Iran's nuclear program emerging as an issue in 2002, the George W. Bush Administration undertook multi-faceted efforts to limit Iran's strategic capabilities through international diplomacy and sanctions. Although Iran appeared to have no role in the September 11, 2001, attacks, President Bush appeared to define Iran as an enemy of the United States when he included Iran as part of an "axis of evil" in his January 2002 State of the Union message (along with Iraq and North Korea). President George W. Bush's second inaugural address (January 20, 2005) and his State of the Union message of January 31, 2006, stated that the United States would be a close ally of a free and democratic Iran, perhaps reflecting the views of those in the Administration, such as Vice President Richard Cheney, who argued that policy should focus on U.S. efforts to change Iran's regime.[64]

Bush Administration statements that it considered Iran a great nation and respects its history[65] reflected the views of those in the Bush Administration who believed in keeping the option of diplomacy open—particularly in light of Iran's potential to harm U.S. troops deployed in Afghanistan and Iraq. The Administration engaged Iran on specific regional issues: for example, it conducted a dialogue in Geneva with Iran on Iraq and Afghanistan from late 2001 until May 2003.[66] This represented the first confirmed direct dialogue between the two countries since the 1979 revolution. The United States aided victims of the December 2003 earthquake in Bam, Iran, including a reported offer—rebuffed by Iran—to send a high-level delegation to Iran, reportedly including then Senator Elizabeth Dole and President George W. Bush's sister, Dorothy. An amendment by then Senator Joseph Biden to the FY2007 defense authorization bill (P.L. 109-364) supported the Administration joining nuclear talks with Iran.

"Grand Bargain Concept"

The George W. Bush Administration did not offer Iran an unconditional, direct U.S.-Iran bilateral dialogue on all issues of U.S. concern. However, some say the Bush Administration "missed an opportunity" for a "grand bargain" with Iran on its nuclear program and other issues of concern by rebuffing a reported comprehensive overture from Iran just before the May 12, 2003, Riyadh bombing. The *Washington Post* reported on February 14, 2007, ("2003 Memo Says Iranian Leaders Backed Talks") that the Swiss ambassador to Iran in 2003, Tim Guldimann, had informed U.S. officials of a comprehensive Iranian proposal for talks with the United States.[67] However, State Department officials and some European diplomats based in Tehran at that time question whether that proposal represented an authoritative Iranian communication. Others argue that the offer was unrealistic because an agreement would have required Iran to abandon key tenets of its Islamic revolution.

Obama Administration Policy: Engagement Coupled with Pressure

President Obama's Administration took office asserting that there was an opportunity to diplomatically dissuade Iran from expanding its nuclear program and potentially to build a new framework for relations with Iran after the decades of estrangement and enmity. After taking office, the Administration offered to integrate Iran into the world economy in return for Iranian compromises on its nuclear program. Some officials, including Secretary of State Clinton and then Secretary of Defense Gates expressed skepticism that engagement would yield changes in Iran's policies. Others believed that the United States needed to present Iran with clear incentives and punishments for continuing uranium enrichment.

The first major public manifestation of President Obama's approach to Iran policy came in his message to the Iranian people on the occasion of Nowruz (Persian New Year), March 21, 2009. He stated that the United States "is now committed to diplomacy that addresses the full range of issues before us, and to pursuing constructive ties among the United States, Iran, and the international community." He also referred to Iran as "The Islamic Republic of Iran," a formulation that suggests an aversion to any regime change option. Prior to the June 12, 2009, election in Iran, other steps included:

- President Obama's reported two letters in 2009 to Iran's Supreme Leader expressing the Administration's philosophy in favor of engagement with Iran.
- A major speech to the "Muslim World" in Cairo on June 4, 2009, in which President Obama said the United States had played a role in the overthrow of Mossadeq, and said that Iran had a right to peaceful nuclear power if it complies with its responsibilities under the NPT.
- An announcement on April 8, 2009, that U.S. officials would attend all P5+1 meetings with Iran.
- Loosened restrictions on U.S. diplomats to meet their Iranian counterparts at international meetings, and permission for U.S. embassies to invite Iranian diplomats to the 2009 celebration of U.S. Independence Day. (The July 4, 2009, invitations did not get issued because of the Iran unrest.)

2010-2012: Focus on Pressure and Preventing a Nuclear-Armed Iran

The crackdown on the 2009 election-related unrest by Iran and its refusal to agree to technical terms of the October 1, 2009, nuclear agreement shifted the Administration's focus to pressuring Iran economically and diplomatically. In a statement following the June 9, 2010, passage of Resolution 1929, President Obama described Iran as refusing to accept the path of engagement and choosing instead to preserve all elements of its nuclear program. Since then, the Administration has emphasized imposing and implementing additional sanctions, while still offering dialogue and negotiations if Iran is willing to bargain seriously on the core concerns about its nuclear program. In 2012, President Obama has explicitly ruled out containing a nuclear Iran and stated that U.S. policy is to prevent Iran from becoming a nuclear state. Secretary of Defense Panetta said on August 1, 2012, during a visit to Israel that "We [the United States] will not allow to develop a nuclear weapon. Period." S.J.Res. 41, discussed below, and which passed the Senate on September 22, 2012, rejects any U.S. policy that relies on containment but acknowledges that President Obama has ruled out relying on containment.

The Administration has downplayed the possibility of military action but stated repeatedly that this U.S. option remains "on the table." In attempting to dissuade Israel from striking Iran militarily, President Obama said during the March 4-6, 2012, visit of Israeli Prime Minister Benjamin Netanyahu that sanctions are beginning to work and should be given more time. President Obama has made this argument several times since the visit, based on signs that Iran's economy is suffering severely and on Iran's willingness to re-engage in nuclear negotiations in 2012. However, amid Israeli statements in August and September 2012 indicating that the talks with Iran have proved fruitless, the White House said on August 21, 2012, that there is still "time and space…to resolve this diplomatically..But it is also clear that the window that is open now to resolve this diplomatically will not remain open indefinitely." President Obama repeated a similar formulation at his first post-re-election press conference on November 14, 2012.

With the nuclear talks not yielding clear progress, the Administration has continued to impose additional sanctions beyond the EU oil embargo and related U.S. sanctions that took full effect as of July 1, 2012. The President signed into law the Iran Threat Reduction and Syria Human Rights Act on August 10, 2012 (P.L. 112-158, H.R. 1905), a bill that, among other provisions, sanctions the transport of oil from Iran.

At the same time, reports began to appear in mid-October 2012 that the Administration might be considering direct bilateral talks with Iran to inject progress into the nuclear talks.[68] Other versions say the Administration might consider direct talks with Iran on regional issues Iran considers significant, such as Bahrain and Syria. Some unconfirmed press reports say there have been talks between U.S. officials and their Iranian counterparts, for example in Turkey or Qatar. The potential willingness of the Administration to undertake such talks - even though the Administration has said no such direct talks are scheduled - has apparently set off a debate inside Iran over the benefits of such talks and of possibly reducing tensions with the United States. Some key leaders, including Ahmadinejad (November 8, 2012) and Mohammad Larijani have backed the idea of direct talks while IRGC and some other hardliners have said the United States needs to change its behavior (ie. lift sanctions) before such talks should be considered. This debate was reflected in a report by Iran's Ministry of Intelligence in November 2012, normally considered a hardline bastion, that highlight the potential benefits of negotiating a solution to the nuclear dispute.

In concert with the democratic uprisings in the Middle East that began in 2011, the Administration also has expressed more direct criticism of Iran for its human rights abuses. As noted above, President Obama's March 20, 2011, Nowruz statement was significantly more supportive of the pro-democracy movement in Iran than it was in prior years. The focus of his March 20, 2012, Nowruz statement was on stating that the United States will seek to help Iranians circumvent government restrictions on the Internet and other media forms. These pronouncements have been supported by imposition of sanctions on Iranian human rights abusers, as discussed further below in the section on regime change.

U.S. and Other Military Action "On The Table" as Israel Threatens a Unilateral Strike

President Obama has repeatedly stated that military options are "on the table." In a March 2, 2012, interview (*The Atlantic*), he clarified that when he says that "all options are on the table" he means that there is a military component to preventing a nuclear-armed Iran.[69] Yet, President Obama and other senior officials continued to maintain that they prefer a diplomatic solution and that military action is a last resort if sanctions and diplomacy fail. And, the Administration has refused to accept Israeli urgings—such as by Prime Minister Netanyahu in his September 27, 2012, U.N. General Assembly speech, to set clear "red lines"—actions that, if taken by Iran, would trigger certain U.S. military action.

Senior U.S. officials have repeatedly stressed the potential adverse consequences of military action, such as Iranian retaliation that might expand throughout the region or even beyond, a reduction of Iran's regional isolation, a strengthening of Iran's regime domestically, an escalation of world oil prices, and the likelihood that military action would only delay Iran's eventual acquisition of a nuclear weapons capability by about one to two years. These points were enumerated by Secretary of Defense Panetta in a speech to the Brookings Institution on December 2, 2011.[70] Most U.S. allies oppose military action, and some reports say that British military leaders might bar the use of British bases for any pre-emptive U.S. strike on Iran.

Some argue that U.S. military action could set back Iran's nuclear program substantially because there are a limited number of key targets, all of which are relatively vulnerable, including the enrichment site at Fordow that is hardened.[71] On the other hand, reports about U.S. confidence in its ability to do substantial damage to any Iranian nuclear target could be intended to signal to Israel—which is threatenening unilateral action as discussed below—that the United States can destroy Iran's nuclear infrastructure, if needed and so ordered.

Iranian Retaliation and Alternative Scenarios of U.S.-Iran Conflict[72]

Consistent with U.S. assessments, Iran's Supreme Leader and other Iranian political and military figures have repeatedly warned that Iran will retaliate for any U.S. military action taken against Iran. In September 2012, IRGC officials warned that even if military action were taken only by Israel, the action would trigger retaliation against U.S. targets. U.S. officials view as increasingly significant the potential for Iran to try to retaliate inside the United States through terrorism. Some believe that Iran would try to retaliate against U.S. personnel abroad, such as embassies and facilities in Europe or the Persian Gulf. Iran could also try to direct anti-U.S. militias in Iraq and Afghanistan to attack U.S. personnel. There are no U.S. troops left in Iraq, but there are about 13,000 U.S. personnel at diplomatic installations there.

Iran's capability to retaliate appears to be increasing. The Defense Department April 2012 report on Iranian military power indicates that Iran's retaliatory ability is growing, as discussed above. That ability has been enhanced through acquisition of additional ships and submarines, increasingly accurate and lethal short range ballistic missiles, and new missile capabilities to re- target, while in flight, U.S. ships or related installations. Iran has also added naval bases along its Gulf coast, according to the DoD report, enhancing its ability to threaten shipping in the Strait. Many U.S. naval planners also worry about Iran's ability to use its large fleet of small boats to "swarm" U.S. ships, and its ability to lay numerous mines in the narrow

Strait of Hormuz. On November 1, 2012, an Iranian Su-25 fired at but missed a U.S. drone that the United States said was over international waters. Earlier, Iran demonstrated its willingness and ability to act militarily: in February 2007, Iran seized 15 British sailors that Iran said were patrolling in Iran's waters, although Britain says they were in Iraqi waters performing coalition-related searches. They were held until April 5, 2007.

To reduce the effectiveness of Iranian retaliation, some argue that the United States would need to strike not only nuclear facilities but all of the retaliatory capabilities discussed above. Earlier, press reports in late February 2012, citing reported Defense Department briefings of President Obama on military options on Iran, say that a U.S. strike could include IRGC and leadership targets.

Still others argue that there are U.S. military options available that do not require actual combat. Some say that a naval embargo or related embargo is possible and could pressure Iran into reconsidering its stand on the nuclear issue. Others say that the imposition of a "no-fly zone" over Iran might also serve that purpose. Still others say that the United Nations could set up a special inspection mission to dismantle Iran's WMD programs, although inserting such a mission is likely to be resisted by Iran and could involve hostilities. A U.S. ground invasion to remove Iran's regime has not, at any time, appeared to be under serious consideration in part because of the likely resistance an invasion would meet in Iran.

U.S.-Iran military conflict could result from events or actions other than a deliberate U.S. decision to strike Iran. Such possibilities drew increased attention in relation to Iran's repeated threats to close the Strait of Hormuz if sanctions are imposed on Iran's exportation of oil. Iran reiterated that threat in late June 2012 on the eve of the entry into full force of the EU oil embargo on Iran on July 1, 2012.

Containment of Iran and U.S. and Gulf Defensive Preparations

Some believe Iran will inevitably become a nuclear armed state, no matter what policies are put into effect, and that containing a nuclear armed Iran is a viable option over the long term. Critics see a reliance on containment as an abandonment of U.S. efforts to prevent Iran from becoming a nuclear state. As noted above, during the visit of Netanyahu in early March 2012, President Obama explicitly ruled out such a strategy.

Several "sense of Congress" resolutions introduced in the 112th Congress urge the President to "reaffirm the unacceptability of an Iran with nuclear weapons capability and oppose any policy that would rely on containment as an option in response to the Iranian nuclear threat." H.Res. 568 passed the House on May 17, 2012, by a vote of 401-11. Its companion measure in the Senate, S.Res. 380 did not advance. On September 22, 2012, the Senate passed S.J.Res. 41, by a vote of 90-1, containing language similar to H.Res. 569 and S.Res. 380, with an additional clause acknowledging that President Obama had ruled out the containment option. Provisions of the National Defense Authorization Act for FY2013 (H.R. 4310) would mandate additional Defense Department planning for exercises and other actions, such as those involving the U.S. Fifth Fleet that is headquartered in Bahrain, that would demonstrate U.S. resolve to Iran.

Even though Iran has not yet acquired a nuclear weapon, many elements of a containment strategy have already been put in place by successive Administrations to try to prevent that outcome, or to limit Iran's regional influence in general. A key component of the strategy has been to enhance the capabilities of U.S. allies in the Persian Gulf. Major initiatives to do so were put in place during the Clinton Administration and further developed during the Bush Administration. In mid-2006 the State Department, primarily the Bureau of Political-Military Affairs ("Pol-Mil"), established the "Gulf Security Dialogue" (GSD). The Obama Administration has continued the GSD effort. During a visit to the Middle East in March 2009, Secretary of State Clinton said, after meeting with several Arab and Israeli leaders in the region, that "there is a great deal of concern about Iran from this whole region." Iran was also the focus of her trip to the Gulf region (Qatar and Saudi Arabia) in February 2010, in which she again raised the issue of a possible U.S. extension of a "security umbrella" or guarantee to regional states against Iran, as a means of preventing Gulf accommodations to Iranian demands or attempting themselves to acquire countervailing nuclear capabilities. The Middle East unrest that spread to the Gulf states of Bahrain and Oman in 2011 caused the Administration to announce a broad arms sales review of all Middle East arms sales. That at first appeared to place the GSD concept in some doubt, but the continuing threat from Iran has largely overtaken those considerations—with the exception of those to Bahrain, most major arms sales to the Gulf states appear to be continuing without interruption.

An cornerstone of the GSD has been to improve and link into one system Gulf state missile defense capabilities, as well as to improve border and maritime security equipment through sales of combat littoral ships, radar systems, and communications gear. During her visit to Saudi Arabia on March 30-31, 2012, Secretary Clinton inaugurated a U.S.-GCC strategic dialogue that revived the longstanding concept of a GCC-wide, integrated missile defense architecture. She again discussed this issue with GCC leaders during a meeting at the margins of the late September 2012 meetings at the U.N. General Assembly. Several GSD-inspired missile defense sales include PAC-3 sales to UAE and Kuwait, and Joint Direct Attack Munitions (JDAMs) to Saudi Arabia and UAE; and the very advanced "THAAD" (Theater High Altitude Area Defense) to UAE. The THAAD sale, previously notified to Congress, was finalized in early January 2012.[73] In early September 2012, it was reported that the United States would soon put in place an early-warning missile defense radar in Qatar that, when combined with radars in Israel and Turkey, would provide a wide range of coverage against Iran's missile forces.[74]

Other Recent and Pending Deployments to Deter Iran and Reassure U.S. Allies

To enhance its strike capabilities and reassure its allies, including Israel, the United States has announced augmented capabilities since June 2012. During that period it has moved a second aircraft carrier into the Gulf region, added a ship (*USS Ponce*) in the Gulf that serves as a platform for helicopters and Special Operations Forces, and added minesweeping capabilities including underwater drones that can find and destroy mines. In late September 2012, the United States and 30 other nations held minesweeping exercises in the Persian Gulf to reinforce their ability to respond to any Iranian military action there. On November 15, 2012, it was reported that the United States was sending additional advanced anti-mine technology to the Gulf.

Other Strategic Missile Defense Concepts against Iran

As part of the effort to demonstrate to Iran that nuclear weapons have no utility, there has also been planning to defend against an eventual long-range Iranian missile system. In August 2008, the George W. Bush Administration reached agreements with Poland and the Czech

Republic to establish a missile defense system to counter Iranian ballistic missiles. These agreements were reached over Russia's opposition, which was based on the belief that the missile defense system would be used to neutralize Russian capabilities. However, reportedly based on assessments of Iran's focus on missiles of regional range, on September 17, 2009, the Obama Administration reoriented this missile defense program to focus, at least initially, on ship-based systems, and systems based in other European countries, including Romania, possibly later returning to the idea of Poland- and Czech-based systems. Some saw this as an effort to win Russia's support for additional sanctions on Iran, although Russia continues to disagree with the plan.

Presidential Authorities and Legislation on Military Action

Perhaps in the belief that there needs to be more advanced planning for U.S. military action, the FY2011 Defense Authorization Act (P.L. 111-383, signed January 7, 2011) contained a provision (§1243) requiring the Administration to develop a "National Military Strategy to Counter Iran." A provision (Section 307) of a FY2013 intelligence authorization bill (H.R. 5743) requires an assessment of the consequences of a military strike on Iran.

A decision to take military action might raise the question of presidential authorities, but no legislation has been passed by both chambers and signed into law limiting the President's authority to use military force against Iran. In the 109th Congress, H.Con.Res. 391, introduced on April 26, 2006, called on the President to not initiate military action against Iran without first obtaining authorization from Congress. A similar bill, H.Con.Res. 33, was introduced in the 110th Congress. An amendment to H.R. 1585, the National Defense Authorization Act for FY2008, requiring authorization for force against Iran, was defeated 136 to 288. A provision that sought to bar the Administration from taking military action against Iran without congressional authorization was taken out of an early draft of an FY2007 supplemental appropriation (H.R. 1591) to fund additional costs for Iraq and Afghanistan combat (vetoed on May 1, 2007). Other provisions, including requiring briefings to Congress about military contingency planning related to Iran's nuclear program, were in a House-passed FY2009 defense authorization bill (H.R. 5658).

Incidents at Sea Agreement?

In the 111th Congress, H.Con.Res. 94 called for the United States to negotiate an "Incidents at Sea" agreement with Iran. Section 1240 of the National Defense Authorization Act for FY2011 (P.L. 111-383) calls for a DOD report, within one year of enactment, on the merits of such an agreement with Iran and other Persian Gulf countries. A press report in September 2011 said that some Defense Department officials are contemplating establishing formal communications channels to Iranian naval officers to prevent misunderstandings and unintended conflict.[75] The idea grew out of a series of incidents with Iranian vessels, some of the incidents involving British warships, that nearly prompted confrontation with Iran.

An Israeli Strike?[76]

Some Israeli leaders, including Prime Minister Benjamin Netanyahu, view a nuclear Iran as a potential existential threat, and say they are determined to prevent a nuclear Iran, even if Israel has to take unilateral action. Implicit in these Israeli statements is the view that Israeli leaders do not completely trust that the United States will take military action to prevent Iran from becoming a nuclear state. Following the November 2011 IAEA report and the start of enrichment activities at the hardened site at Fordow, Defense Minister Ehud Barak and other Israeli leaders expressed concern that Iran's nuclear program might be entering a "zone of immunity" beyond which Israel will have no military options.

In his U.S. visit during March 4-5, 2012, Prime Minister Netanyahu expressed the right of Israel to act in its self-defense. President Obama, in speeches to the America-Israel Public Affairs Committee on March 4, 2012, and in statements before and after the Netanyahu visit, acknowledged Israel's "sovereign right" of self-defense.[77] While the possibility of an Israeli strike on Iran is widely viewed as low while the Iran-P5+1 negotiating process continues, Israeli leaders reportedly have again stepped up internal discussion of this option as the prospects for success in the P5+1—Iran talks has faded. Several high level U.S. visits to Israel in July 2012, including by National Security Advisor Donilon, Secretary Clinton, and Secretary Panetta, were apparently intended to ensure full U.S.-Israel coordination on policy toward Iran by assuring Israel that the United States is fully committed to preventing a nuclear-armed Iran. And, U.S. deployments in the Gulf, discussed above, as well as additional planned moves to implement sanctions, and possible planned covert action, have been intended to reassure Israel that the United States is dealing adequately with the nuclear threat from Iran.[78] Still, in his September 27, 2012, U.N. General Assembly speech, Netanyahu stressed the need for the international community, primarily the United States, to set a clear red line for Iran short of acquiring enough high enriched uranium for a nuclear weapon. The Administration efforts might be succeeding because, as of November 2012, Israeli officials have been stressing imposing additional Iran sanctions rather than military options and pushing out the time horizon for possible unilateral Israeli military action until at least early 2013. On November 9, 2012, Defense Minister Barak appeared to push that horizon out further to "eight to ten months," taking it to mid-2013. This shift may have been caused by an Israeli assessment that it not injure relations with the Obama Administration over the issue of taking military action, as well as by evidence that sanctions are weakening the Iranian economy are mounting.

Although Israeli strategists say that a strike might be a viable option, several U.S. experts doubt that Israel has the capability to make such action sufficiently effective to justify the risks. The IAF is capable but far smaller than that of the United States, and could require overflight of several countries not likely to support Israeli action, such as Iraq.

While the Israeli strike possibility has become acute, it is not new. In mid-June 2008, Israeli officials confirmed reports that the Israel Air Force (IAF) had practiced a long-range strike such as that which would be required for an attack on Iran's nuclear sites. Debate recurred in September 2010 following the publication of an article in *The Atlantic* by Jeffrey Goldberg entitled "Point of No Return" that hinted at a possibly impending strike in early 2011.[79]

Reported Covert Action

As international concern about Iran's nuclear program has grown, there is increasing discussion about a reported covert component to U.S. attempts to slow Iran's nuclear program. An option is for the United States and partner countries to increase this activity, which is distinct from covert action to support groups inside Iran looking to overthrow Iran's regime.

Previously, during 2006-2008, it was reported that the United States and Israel conducted operations that resulted in the sale to Iran of nuclear and other technology rigged to have a destructive effect on Iran's programs. Another example includes the Stuxnet virus, discussed above. The killings of some Iranian scientists over the past few years remain unexplained and could have been the result of covert action. The latest Iranian scientist to be killed was Mostafa Ahmadi Roshan, a chemical engineer at the Natanz enrichment facility, who died when a bomb placed under his car exploded on January 10, 2012. Earlier, on December 5, 2011, a U.S. drone, the RQ-170 Sentinel, went down in Iran; it reportedly was based in Afghanistan and may have been sent over Iran to monitor Iran's nuclear sites. Iran has refused a U.S. request to return the drone.

Regime Change

Throughout its first year, the Obama Administration sought to allay Iran's long-standing suspicions that the main U.S. goal is to unseat the Islamic regime in Iran. Iran's suspicions of U.S. intentions are based on the widespread perception that the United States has hoped for and at times sought to promote regime change in Iran since the 1979 Islamic revolution. The United States provided some funding to anti-regime groups, mainly pro-monarchists, during the 1980s,[80] and the George W. Bush Administration expressed attraction to this option on several occasions. The Obama Administration's stated policy remains to alter Iran's behavior, not change its regime.

The 2009 domestic unrest in Iran complicated policy for the Obama Administration because it demonstrated that the regime in Iran might be vulnerable to overthrow. However, the Administration, assessing that outcome as unlikely, sought to preserve the possibility of a nuclear agreement with Iran while expressing support for human and political rights demanded by the Green Movement. As 2009 progressed, the statements of President Obama and other U.S. officials became progressively more critical of the regime. On December 28, 2009, President Obama expressed forthright support for the opposition by saying, in regard to the unrest in Iran, "Along with all free nations, the United States stands with those who seek their universal rights."[81] With the protests absent in Iran for nearly a year, Secretary of State Clinton reiterated this position on September 19, 2010, but said the United States needs take care not to be so overtly supportive as to make the Iranian opposition appear as "stooges" of the United States.

In 2011, the Administration reevaluated its stance slightly in the context of the broader Middle East uprisings. Statements by Secretary Clinton and the National Security Council accused Iran of hypocrisy for supporting demonstrations in Egypt while preventing similar free expression inside Iran.[82] Many observers noted that President Obama's 2011 Nowruz address (delivered March 20, 2011, the eve of Nowruz) was far more explicitly supportive of the Iranian opposition than in past years, mentioning specific dissidents who have been jailed and saying to the "young people of Iran ... I want you to know that I am with you."[83] Since that statement, the Administration has, as noted below, sanctioned Iranian officials for human rights abuses in Iran and for assisting Syria with its crackdown against demonstrations. In his speech to the U.N. General Assembly on September 21, 2011, President Obama said "In Iran, we've seen a government that refuses to recognize the rights of its own people." These statements and steps stop short of constituting a policy of "regime change," although Iran interprets any public support for the domestic opposition as evidence of U.S. intent to overthrow the clerical government. As noted above, his 2012 Nowruz message (March 20, 2012) focused on U.S. efforts to help Iranians circumvent government restrictions on the Internet and other media—a so-called "electronic curtain."

Some in Congress appear to advocate more direct, public, and broad U.S. support for the overthrow of the regime as a focus of U.S. policy. In the 111th Congress, one bill said that it should be U.S. policy to promote the overthrow of the regime (The Iran Democratic Transition Act, S. 3008).

Pursuing a Middle Ground: Democracy Promotion and Internet Freedom Efforts

In the absence of all-out U.S. pursuit of regime change, successive Administrations and Congress have agreed on more modest steps to promote political evolution in Iran through "democracy promotion" and sanctioning Iranian human rights abusers.

Sanctioning Iranian Human Rights Abusers and Abuses

As part of its efforts to isolate the regime on human rights grounds, on September 29, 2010, President Obama, acting in accordance with Section 105 of P.L. 111-195 (CISADA), issued Executive Order 13553, imposing sanctions on Iranian officials determined to have committed human rights abuses since Iran's 2009 election. Sanctions include a ban on visas to the United States and freeze on U.S.-based assets or trade with them. In an annex, eight Iranian officials were named as violators and were subjected to the sanctions. Two more Iranian officials (Tehran prosecutor Abbas Dowlatabadi and Basij commander Mohammad Reza Naqdi) were added to that list on February 23, 2011, and, on June 9, 2011, the Administration added the IRGC (already sanctioned as a proliferator), the Basij, the Law Enforcement Forces (LEF), and LEF Commander Ismail Ahmadi Moghaddam to the list.

In the 112th Congress, several bills have been introduced to increase sanctions on Iranian human rights abusers. S. 879 and H.R. 1714 made mandatory investigations of Iranian human rights abusers and sanction the sale to Iran of equipment that could be used to suppress demonstrations. Elements of these several bills were incorporated into a broad Iran sanctions bill, H.R. 1905, passed by both chambers on August 1, 2012, and signed on August 10 (P.L. 112-158).

Democracy Promotion Funding

Binding legislation to favor democracy promotion in Iran was enacted in the 109th Congress. The Iran Freedom Support Act (P.L. 109-293), signed September 30, 2006, authorized funds (no specific dollar amount) for Iran democracy promotion.[84] Iran asserts that funding democracy promotion represents a violation of the 1981 "Algiers Accords" that settled the Iran hostage crisis and provide for non-interference in each others' internal affairs.

The George W. Bush Administration asserted that open funding of Iranian pro-democracy activists (see below) as a stated effort to change regime behavior, not to overthrow the regime, although some saw the Bush Administration's efforts as a cover to achieve a regime change objective. A few accounts, such as "Preparing the Battlefield" by Seymour Hersh in the *New Yorker* (July 7 and 14, 2008) say that President George W. Bush authorized U.S. covert operations to destabilize the regime,[85] involving assistance to some of the ethnic-based armed groups discussed above. CRS has no way to confirm assertions in the Hersh article that up to $400 million was appropriated and/or used to aid the groups mentioned.

The State Department, the implementer of U.S. democracy promotion programs for Iran, has used funds in appropriations (see **Table 9**) to support pro-democracy programs run by at organizations based in the United States and in Europe; the department refuses to name grantees for security reasons. The funds shown below have been obligated through DRL and the Bureau of Near Eastern Affairs in partnership with USAID. At least $60 billion of the funds have been allocated to date. Some of the funds have been appropriated for cultural exchanges, public diplomacy, and broadcasting to Iran. A further indication of the sensitivity of specifying the use of the funds is that the Obama Administration requested funds for Near East regional democracy programs in its FY2010, FY2011, FY2012, and FY2013 budget requests, but no specific requests for funds for Iran were delineated.

Many have consistently questioned the effectiveness of such funding. In the view of many experts, U.S. funds would make the aid recipients less attractive to most Iranians. Even before the post-2009 election crackdown, Iran was arresting civil society activists by alleging they are accepting the U.S. democracy promotion funds, while others have refused to participate in U.S.- funded programs, fearing arrest.[86] In May 2007—Iranian American scholar Haleh Esfandiari, of the Woodrow Wilson Center in Washington, DC, was imprisoned for several months, on the grounds that the Wilson Center was part of this effort. The center has denied being part of the democracy promotion effort in Iran.

Perhaps in response to some of these criticisms, the Obama Administration altered Iran democracy promotion programs somewhat toward working directly with Iranians inside Iran who are organized around such apolitical issues as health care, the environment, and science.[87] During 2009, less emphasis was placed on funding journalists and human rights activists in Iran, or on sponsoring visits by Iranians to the United States.[88] One issue arose concerning the State Department decision in late 2009 not to renew a contract to the Iran Human Rights Documentation Center (IHRDC), based at Yale University, which was cataloguing human rights abuses in Iran. Some outside experts believe that, particularly in the current context of a regime crackdown against democracy activists, the contract should have been renewed. That criticism went hand in hand with the view of some experts that the post-election unrest in Iran was evidence that such democracy promotion programs were working and should be enhanced.

Promoting Internet Freedom in Iran

In line with legislation and new assessments of the best use of U.S. assistance, recent U.S. actions have focused on preventing the Iranian government's suppression of electronic communication. Among legislation that was enacted is the "Voice (Victims of Iranian Censorship) Act" (Subtitle D of the FY2010 Defense Authorization, P.L. 111-84), which contains provisions to potentially penalize companies that are selling Iran technology equipment that it can use to suppress or monitor the Internet usage of Iranians.[89] In February 2010, the Administration eased licensing requirements for Iranians to download free mass market U.S. software. And, the U.S. Office of Foreign Assets Control has reportedly licensed a California firm (Censorship Research Center) to export anti-filtering software to Iran.[90] Under Secretary of State Sherman testified on October 14, 2011, that some of the democracy promotion funding has been to train Iranians in the use of technologies that undermine regime Internet censorship efforts.

In March 2012, the Administration focused on this issue anew. In his March 20, 2012, Nowruz message, President Obama stated that in recent weeks the regime had increased

Internet restrictions and that the Administration is taking new steps to promote Internet freedom in Iran. Acting in accordance with P.L. 111-195 (Comprehensive Iran Sanctions, Accountability, and Divestment Act), which allows U.S. exports of technologies used to expand Internet freedom in Iran, on March 20, 2012, the Administration announced that certain software that can be used to circumvent regime restrictions on Internet use could be exported to Iran without a specific license. The Administration announced examples such as software for personal communications, data storage, Internet browsing, document reading, and related technologies. This appeared to be a major new step in the Administration effort to break Iran's "electronic curtain."

On April 23, 2012, President Obama issued an Executive Order (13606) directly addressing the issue. The order blocks the U.S.-based property and essentially bars U.S. entry and bans any U.S. trade with persons and entities listed in an Annex and persons or entities subsequently determined to be operating any technology that allows the Iranian (or Syrian) government to disrupt, monitor, or track computer usage by citizens of those countries; or to have sold to Iran or Syria any technology that enables those government to carry out such disruptions or monitoring.

Broadcasting/Public Diplomacy Issues

Another part of the democracy promotion effort has been the development of new U.S. broadcasting services to Iran. The broadcasting component of policy has been an extension of a trend that began in the late 1990s. Radio *Farda* ("tomorrow," in Farsi) began under Radio Free Europe/Radio Liberty (RFE/RL), in partnership with the Voice of America (VOA), in 2002. The service was established as a successor to a smaller Iran broadcasting effort begun with an initial $4 million from the FY1998 Commerce/State/Justice appropriation (P.L. 105-119). It was to be called Radio Free Iran but was never formally given that name by RFE/RL. Radio *Farda* now broadcasts 24 hours/day. Radio *Farda* has 59 full time employees. No U.S. assistance has been provided to Iranian exile-run stations.[91]

According to information provided to CRS by the BBG, the costs of Radio Farda are FY2010: $9.9 million; FY2011: $11.84 million; and FY2012: $11.77 million.

Persian News Network (PNN). The VOA established a Persian language service to Iran (VOA Persian Service) in July 2003. In July 2007, it was renamed Persian News Network (PNN), encompassing radio (1 hour a day or original programming); television (6 hours a day of primetime programming, rebroadcast throughout a 24-hour period); and Internet.

Even though PNN has expanded its offerings significantly, it has come under substantial criticism from observers. Some say that PNN has lost much of its audience among young, educated, anti- regime Iranians who are looking for signs of U.S. official support. The Inspector General report cited above, as well as many observers maintain that decisions on who to put on PNN panel discussion shows have been made by a small group of Iranian exiles who deliberately deny appearances to certain Iranians with whom they disagree. Still others say that PNN frequently airs the views of Iranian groups that are advocates of U.S. engagement of the regime or who downplay regime transgressions. Some have criticized PNN for covering long-standing exiled opposition groups, such as supporters of the son of the former Shah of Iran.[92] Other critics say PNN offers little coverage of the Green Movement, even though a mission of the network is, or should be, to highlight the purported unpopularity of the regime. Others say it has run pieces pointing out such U.S. social problems as homelessness and drug use, while refusing to air pieces showcasing U.S. democracy and rule of law. Other observers say there is wide-ranging nepotism at PNN, in which employees hire their relatives and deny opportunities to employment applicants based on merit. VOA officials said in September 2012 they are attempting address some of these issues through the human resources office of the VOA.

Several observers point to one particular PNN show as having particular effect on audiences inside Iran. That show is called "Parazit" (Persian for static); it is a weekly comedy show modeled on a U.S. program on Comedy Central network called "The Daily Show." On Parazit, the writers of the show, Kambiz Hosseini and Saman Arbabi, mocked President Mahmoud Ahmadinejad and other Iranian figures, using political satire. Observers say that the show deteriorated in quality in 2012 after Mr. Hosseini left the show or was ousted from it, and it was taken off PNN in February 2012. A different show that satirizes Iranian leaders and news from Iran—called On Ten—began in April 2012, and VOA officials say it is gaining a substantial audience. PNN said in August 2012 that Parazit is to be reconstituted, and based in New York.

Other issues relate to the PNN's leadership and governance. In February 2011, Ramin Asgard, a former State Department officer, was hired as PNN director.[93] tasked with redressing the PNN deficiencies. However, he left in January 2012, reportedly out of frustration at his inability to restructure PNN and make it more effective as a voice for U.S. policy. PNN is now temporarily run by VOA official Steve Redisch. He is assisted, in part, by former CNN anchor Rudi Bakhtiar, mentioned above, although she previously was briefly ousted by PNN under unclear circumstances. The VOA is attempting to recruit a permanent replacement for Asgard, and hire a deputy director and an executive producer.

According to a VOA briefing on September 21, 2012, PNN has 79 full-time employees and 114 contractors. Past costs for PNN are: FY2010, $23.78 million; FY2011, $22.5 million; and FY2012 (estimate), $23.32 million. In FY2013 its costs are expected to be about $18 million.

State Department Diplomatic and Public Diplomacy Efforts

Since 2006, the State Department has been increasing the presence of Persian-speaking U.S. diplomats in U.S. diplomatic missions around Iran, in part to help identify and facilitate Iranian participate in U.S. democracy-promotion programs. The Iran unit at the U.S. consulate in Dubai has been enlarged significantly into a "regional presence" office, and "Iran-watcher" positions have been added to U.S. diplomatic facilities in Baku, Azerbaijan; Istanbul, Turkey; Frankfurt, Germany; London; and Ashkabad, Turkmenistan, all of which have large expatriate Iranian populations and/or proximity to Iran.[94] An enlarged (eight-person) "Office of Iran Affairs" has been formed at the State Department, and it is reportedly engaged in contacts with U.S.-based exile groups such as those discussed earlier.

The State Department also is trying to enhance its public diplomacy to reach out to the Iranian population.

- In May 2003, the State Department added a Persian-language website to its list of foreign language websites, under the authority of the Bureau of International Information Programs. The website, according to a statement issued by then-Secretary of State Colin Powell, is intended to be a source of information about the United States and its policy toward Iran.

- On February 14, 2011, the State Department announced that it had begun Persian-language Twitter feeds in an effort to connect better with Internet users in Iran.
- In part to augment U.S. public diplomacy, the State Department announced in April 2011 that a Persian-speaking U.S. diplomat, Alan Eyre, based at the U.S. Consulate in Dubai, would make regular appearances on Iranian official media to explain U.S. positions.
- On October 27, 2011, Secretary Clinton announced the United States would set up a "virtual Embassy" to Iran on the Internet, which would provide Iranians with information on visas to the United States and exchange programs.

Table 9. Iran Democracy Promotion Funding

FY2004	Foreign operations appropriation (P.L. 108-199) earmarked $1.5 million for "educational, humanitarian and non-governmental organizations and individuals inside Iran to support the advancement of democracy and human rights in Iran." The State Department Bureau of Democracy and Labor (DRL) gave $1 million to a unit of Yale University, and $500,000 to National Endowment for Democracy.
FY2005	$3 million from FY2005 foreign aid appropriation (P.L. 108-447) for democracy promotion. Priority areas: political party development, media, labor rights, civil society promotion, and human rights.
FY2006	$11.15 for democracy promotion from regular FY2006 foreign aid appropriation (P.L. 109-102). $4.15 million administered by DRL and $7 million for the Bureau of Near Eastern Affairs.
FY2006 supp.	Total of $66.1 million (of $75 million requested) from FY2006 supplemental (P.L. 109-234): $20 million for democracy promotion; $5 million for public diplomacy directed at the Iranian population; $5 million for cultural exchanges; and $36.1 million for Voice of America-TV and "*Radio Farda*" broadcasting. Broadcasting funds are provided through the Broadcasting Board of Governors.
FY2007	FY2007 continuing resolution provided $6.55 million for Iran (and Syria) to be administered through DRL. $3.04 million was used for Iran. No funds were requested.
FY2008	$60 million (of $75 million requested) is contained in Consolidated Appropriation (H.R. 2764, P.L. 110- 161), of which, according to the conference report $21.6 million is ESF for pro-democracy programs, including non-violent efforts to oppose Iran's meddling in other countries. $7.9 million is from a "Democracy Fund" for use by DRL. The Appropriation also fully funded additional $33.6 million requested for Iran broadcasting: $20 million for VOA Persian service; and $8.1 million for Radio Farda; and $5.5 million for exchanges with Iran.
FY2009	Request was for $65 million in ESF "to support the aspirations of the Iranian people for a democratic and open society by promoting civil society, civic participation, media freedom, and freedom of information." H.R. 1105 (P.L. 111-8) provides $25 million for democracy promotion programs in the region, including in Iran.
FY2010	$40 million requested and used for Near East Regional Democracy programming. Programs to promote human rights, civil society, and public diplomacy in Iran constitute a significant use of these region-wide funds.
FY2011	$40 million requested and will be used for Near East Regional Democracy programs. Programming for Iran with these funds to be similar to FY2010
FY2012	$35 million for Near East Regional Democracy, and Iran-related use is to be similar to FY2010 and FY2011.
FY2013	$30 million for Near East Regional Democracy, with Iran use similar to prior two fiscal years.

Sources: Information provided by State Department and reviewed by Department's Iran Office, February 1, 2010; FY2012 Congressional Budget Justification; author conversation with Department of State Iran Office, April 21, 2011.

Option: Enhanced U.S. Interests Section

Some go further and say that the United States should staff the U.S. interests section in Tehran with U.S. personnel, who would mostly process Iranian visas and help facilitate U.S.-Iran people- to-people contacts (the interests section is currently under the auspices of the Swiss Embassy). U.S. staffing was considered by the George W. Bush Administration in late 2008, but the decision was left to the Obama Administration. The Obama Administration appeared inclined toward U.S. staffing, but no decision was announced. Such a step was likely delayed or derailed outright by the Iranian response to the post-election protests. However, some observers say that there are State Department officials who see U.S. staffing as a way to broaden U.S. contacts with representatives of the Green Movement and more accurately gauge its strength. Perhaps as a temporary alternative, the State Department is attempting outreach to the Iranian people by establishing, as of November 2011, an Internet-based "virtual embassy," that explains the visa application process and other items of interest to Iranians. However, press reports say Iran has censored the site and rendered it at least partially inaccessible.

Additional Sanctions

Amid signs that sanctions are weakening Iran's economy and possibly pressuring its leadership, the Administration and its international partners continue to impose additional sanctions on Iran—particularly with nuclear talks failing to achieve a breakthrough after three rounds during April—June 2012. The lack of clear progress in the talks caused a major Iran sanctions bill, H.R. 1905, the Iran Threat Reduction and Syria Human Rights Act of 2012 to advance. Different versions of the bill, each of which has multiple provisions, were passed by both chambers and an agreed version—which has provisions not in either prior version—was filed on July 30, 2012, and passed by both chambers on August 1, 2012. It was signed on August 10 (P.L. 112-158). The primary new provision in the enacted version is to sanction shipments of Iranian oil unless the country buying the Iranian oil has received an exemption from sanctions for significantly reducing purchases of oil from Iran each 180 days. Thus far, 20 countries—all of Iran's main oil customers—have reduced their buys of Iranian oil and received such an exemption. This provision is highly similar to an Executive Order 13622 issued on July 31, 2012. On October 15, 2012, the EU imposed additional sanctions on Iran, including a ban on imports of natural gas from Iran, which reach several EU states via Turkey.

Aside from the provisions of H.R. 1905, there are numerous ideas and suggestions for additional economic and diplomatic sanctions against Iran, should the negotiating process break down completely. Some are U.S. sanctions, some are U.S. sanctions against foreign entities intended to compel them to exit the Iranian market, and others are multilateral or international.

- *Mandating Reductions in Diplomatic Exchanges with Iran or Prohibiting Travel by Iranian Officials.*
- *Banning Passenger Flights to and from Iran.*
- *Limiting Lending to Iran by International Financial Institutions.* Resolution 1747 calls for restraint on but does not outright ban international lending to Iran.

- *Banning Trade Financing or Official Insurance for Trade Financing.* This was not made mandatory by Resolution 1929, but several countries imposed this sanction (as far as most trade financing) subsequently.
- *Banning Investment in Iran's Energy Sector.* Such a step is authorized, not mandated, by Resolution 1929, and the Iran Sanctions Act allows for U.S. sanctions against foreign investment in that sector. A growing number of countries have used that authority to impose these sanctions on Iran.
- *Restricting Operations of and Insurance for Iranian Shipping.* A call for restraint is in Resolution 1929, but is not mandatory. The EU and other national measures announced subsequently did include this sanction (IRISL) to take effect as of July 1.
- *Imposing a Worldwide Ban on Sales of Arms to Iran.* Resolution 1929 imposes a ban on sales of major weapons systems to Iran, but another option is to extend that ban to all lethal equipment.
- *Imposing an International Ban on Trade With Iran, Particularly Purchases of Iranian Oil.* As noted, the EU has agreed to stop all purchases of Iranian oil as of July 1, 2012, and other countries have cut their buys. An option is to impose a worldwide ban on all purchases, or to further pressure or incent nations to end buys of oil from Iran. A related idea could be the enactment of a global ban on trade with Iran or of U.S. sanctions that seek to compel a partial or comprehensive global ban on trade with Iran.

Table 10. Digest of Existing U.S. Sanctions Against Iran

Ban on U.S. Trade With and Investment in Iran. Executive Order 12959 (May 6, 1995) bans almost all U.S. trade with and investment in Iran. Modifications in 1999 and 2000 allowed for exportation of U.S. food and medical equipment, and importation from Iran of luxury goods (carpets, caviar, dried fruits, nuts), but P.L. 111-195 (Comprehensive Iran Sanctions, Accountability, and Divestment Act, CISADA) restored the complete ban on imports. The trade ban does not apply to foreign subsidiaries of U.S. firms.
U.S. Sanctions Against Foreign Firms that Deal With Iran's Energy Sector. The Iran Sanctions Act (P.L. 104-172, August 5, 1996, as amended, most recently by H.R. 1905/P.L. 112-158) authorizes the President to select five out of a menu of twelve sanctions to impose against firms that the Administration has determined: have invested more than $20 million to develop Iran's petroleum (oil and gas) sector; that buy Iranian oil (unless such country has a sanctions exemption under the P.L. 112-81, see below); have sold Iran more than $1 million worth of gasoline or equipment to import gasoline or refine oil into gasoline; have sold energy $1 million or more worth of energy equipment to Iran; that provided shipping services to transport oil from Iran; that have engaged in an energy joint venture with Iran outside Iran; or that buy Iran's sovereign debt.
Sanctions On Iran's Central Bank. CISADA bans accounts with banks that do business with the Revolutionary Guard and sanctioned entities and the Treasury Dept. in November 2011 declared Iran's financial system an entity of primary money laundering concern. Section 1245 P.L. 112-81 signed December 31, 2011, prevents new foreign banks that do business with Iran's Central Bank from opening U.S. accounts.
Terrorism List Designation Sanctions. Iran's designation by the Secretary of State as a "state sponsor of terrorism" (January 19, 1984—commonly referred to as the "terrorism list") triggers several sanctions, including the following: (1) a ban on the provision of U.S. foreign assistance to Iran under Section 620A of the Foreign Assistance Act; (2) a ban on arms exports

to Iran under Section 40 of the Arms Export Control Act (P.L. 95-92, as amended); (3) under Section 6(j) of the Export Administration Act (P.L. 96-72, as amended), a significant restriction—amended by other laws to a "presumption of denial"—on U.S. exports to Iran of items that could have military applications; (4) under Section 327 of the Anti-Terrorism and Effective Death Penalty Act (P.L. 104-132, April 24, 1996), a requirement that U.S. representatives to international financial institutions vote against international loans to terrorism list states.
Sanctions Against Foreign Firms that Aid Iran's Weapons of Mass Destruction Programs. The Iran-Syria-North Korea Nonproliferation Act (P.L. 106-178, March 14, 2000, as amended) authorizes the Administration to impose sanctions on foreign persons
or firms determined to have provided assistance to Iran's weapons of mass destruction (WMD) programs. Sanctions include restrictions on U.S. trade with the sanctioned entity.
Sanctions Against Foreign Firms that Sell Advanced Arms to Iran. The Iran-Iraq Arms Nonproliferation Act (P.L. 102-484, October 23, 1992, as amended) provides for U.S. sanctions against foreign firms that sell Iran "destabilizing numbers and types of conventional weapons" or WMD technology.
Ban on Transactions With Foreign Entities Determined to Be Supporting International Terrorism. Executive Order 13324 (September 23, 2001) authorizes a ban on U.S. transactions with entities determined to be supporting international terrorism. The Order was not specific to Iran, coming 12 days after the September 11, 2001, attacks, but several Iranian entities have been designated.
Ban on Transactions With Foreign Entities that Support Proliferation. Executive Order 13382 (June 28, 2005) amended previous executive orders to provide for a ban on U.S. transactions with entities determined to be supporting international proliferation. As is the case for Executive Order 13324, mentioned above, Executive Order 13382 was not specific to Iran. However, numerous Iranian entities, including the IRGC itself, have been designated.
Divestment. A Title in P.L. 111-195 authorizes and protects from lawsuits various investment managers who divest from shares of firms that conduct sanctionable business with Iran.
Counter-Narcotics. In February 1987, Iran was first designated as a state that failed to cooperate with U.S. anti-drug efforts or take adequate steps to control narcotics production or trafficking. The Clinton Administration, on December 7, 1998, removed Iran from the U.S. list of major drug producing countries. This exempted Iran from the annual certification process that kept drug-related U.S. sanctions in place on Iran.
U.S.-Iran Assets Disputes. Iranian leaders continue to assert that the United States is holding Iranian assets, and that this is an impediment to improved relations. See CRS Report RS20871, *Iran Sanctions*, by Kenneth Katzman.
Sanctions Against Human Rights Abuses and Internet Monitoring. Various laws discussed above, and Executive Orders, impose sanctions on named Iranian human rights abusers, and on firms that sell equipment Iran can use to monitor the Internet usage of citizens or employ against demonstrators.

Source: CRS. For analysis and extended discussion of U.S. and international sanctions against Iran, see CRS Report RS20871, *Iran Sanctions*, by Kenneth Katzman.

CONCLUSION

Mistrust between the United States and Iran's Islamic regime has run deep for more than three decades and will be difficult to reverse. Some argue that, no matter who is in power in Tehran, the United States and Iran have a common long-term interest in stability in the

Persian Gulf and South Asia regions. According to this view, major diplomatic overtures toward the regime might not only help resolve the nuclear issue but yield fruit in producing a new, constructive U.S.-Iran relationship.

Others argue that U.S. concerns stem first and foremost from the character of Iran's regime, and that no diplomatic breakthrough is possible until the regime changes. Those who take this view see in the Green Movement the potential to replace the regime and to integrate Iran into a pro- U.S. strategic architecture in the region. Many argue that a wholesale replacement of the current regime could produce major strategic benefits beyond potentially reducing the threat from Iran's nuclear program, including an end to Iran's effort to obstruct a broad Arab-Israeli peace.

Others argue that many Iranians are united on major national security issues and that a new regime would not necessarily align with the United States. Some believe that many Iranians fear that alignment with the United States would produce a degree of U.S. control and infuse Iran with Western culture that many Iranians find un-Islamic and objectionable.

Table 11. Selected Economic Indicators

Economic Growth	2.5% (2011 est.); 3.2% (2010)
Per Capita Income	$12,200/yr purchasing power parity (2011)
GDP	$930 billion purchasing power parity (2011)
Proven Oil Reserves	135 billion barrels (highest after Russia and Canada)
Oil Production/ Exports	About 3.9 million barrels per day (mbd)/ 2.4 mbd exports.
Major Oil/Gas Customers	Remaining customers: primarily China, India, South Korea, Japan, and Turkey. Turkey also buys 8.6 billion cubic meters/yr of gas from Iran. Oil exports have fallen to just over 1 million barrels per day as of September 2012.
Major Export Markets	Mirrors major oil customers.
Major Imports	Germany ($5.6 billion); China ($5 billion); UAE ($4 billion); S. Korea ($2.9 billion); France ($2.6 billion);Italy ($2.5 billion); Russia ($1.7 billion); India ($1.6 billion); Brazil ($1.3 billion); Japan ($1.3 billion).
Major Non-Oil Investments	Renault (France) and Mercedes (Germany)—automobile production in Karaj, Iran— valued at $370 million; Renault (France), Peugeot (France) and Volkswagen (Germany)— auto parts production; Turkey—Tehran airport, hotels; China—shipbuilding on Qeshm Island, aluminum factory in Shirvan, cement plant in Hamadan; UAE financing Esfahan Stee Company; India—steel plant, petrochemical plant; S. Korea—steel plant in Kerman Province; S. Korea and Germany—$1.7 billion to expand Esfahan refinery.
Development Assistance Received	2003 (latest available): $136 million grant aid. Biggest donors: Germany ($38 million); Japan ($17 million); France ($9 million).
Inflation	22.5% (2011)
Unemployment Rate	15.3% (2011)

Sources: CIA, *The World Factbook*; various press; IMF; Iran Trade Planning Division; CRS conversations with experts and foreign diplomats.

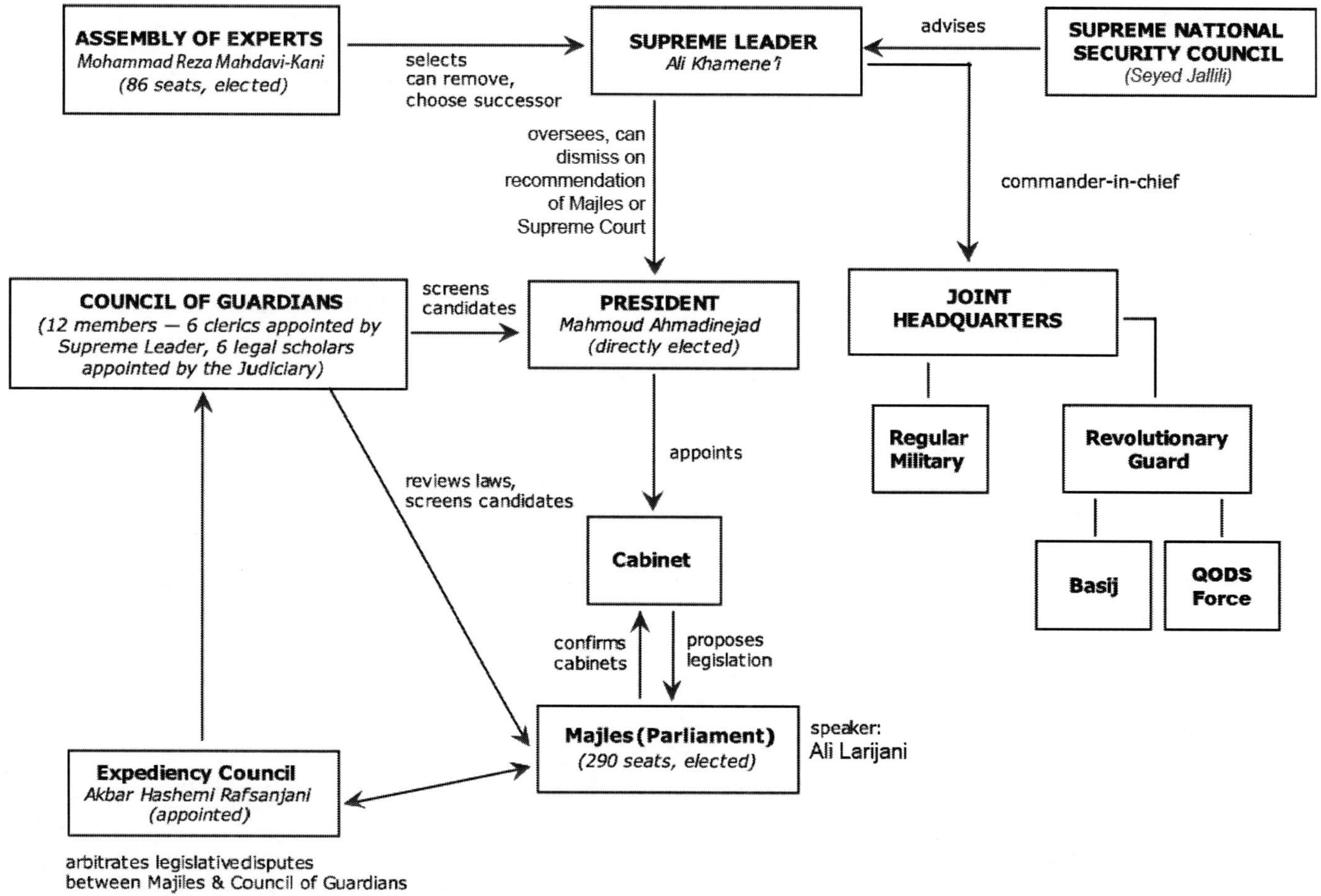

Source: CRS.

Figure 1. Structure of the Iranian Government.

Source: Map Resources, adapted by CRS (April 2005).

Figure 2. Map of Iran.

End Notes

[1] At the time of his selection as Supreme Leader, Khamene'i was generally referred to at the rank of Hojjat ol-Islam, one rank below Ayatollah, suggesting his religious elevation was political rather than through traditional mechanisms.

[2] The Council of Guardians consists of six Islamic jurists and six secular lawyers. The six Islamic jurists are appointed by the Supreme Leader. The six lawyers on the Council are selected by the judiciary but confirmed by the Majles.

[3] Rafsanjani was constitutionally permitted to run because a third term would not have been consecutive with his previous two terms. In the 2001 presidential election, the Council permitted 10 out of the 814 registered candidates.

[4] A paper published by Chatham House and the University of St. Andrews strongly questions how Ahmadinejad's vote could have been as large as reported by official results, in light of past voting patterns throughout Iran. "Preliminary Analysis of the Voting Figures in Iran's 2009 Presidential Election." http://www.chathamhouse.org.uk.

[5] Kampeas, Ron. "Iran's Crown Prince Plots Nonviolent Insurrection from Suburban Washington." Associated Press, August 26, 2002.

[6] Other names by which this group is known is the Mojahedin-e-Khalq Organization (MEK or MKO) and the National Council of Resistance (NCR).

[7] The designation was made under the authority of the Anti-Terrorism and Effective Death Penalty Act of 1996 (P.L. 104-132).

[8] http://www.state.gov/r/pa/prs/ps/2011/04/160404.htm.

[9] Text is at http://www.state.gov/j/drl/rls/hrrpt/humanrightsreport/index.htm?dynamic_load_id= 186425#wrapper.

[10] Human Rights Watch. "Codifying Repression." August 29, 2012.

[11] Department of Defense. Annual Report on Military Power of Iran. April 2012. For the 2010 report, see http://media.washingtontimes.com/media/docs/2010/Apr/20/Iran_Military___Report.pdf. The reports are required by §1245 of the National Defense Authorization Act for FY2010 (P.L. 111-84).

[12] For a more extensive discussion of the IRGC, see Katzman, Kenneth. "The Warriors of Islam: Iran's Revolutionary Guard," *Westview Press*, 1993.

[13] In November 2006, the IAEA, at U.S. urging, declined to provide technical assistance to the Arak facility on the grounds that it was likely for proliferation purposes.

[14] Text of the report is at http://isis-online.org/uploads/isis-reports/documents/IAEA_Iran_8Nov 2011. pdf.

[15] http://www.isis-online.org/uploads/isis-reports/documents/Iran_report-nov23.pdf; http://lewis.armscontrolwonk.com/files/2011/02/gov2011-7.pdf.

[16] For Iran's arguments about its program, see Iranian paid advertisement "An Unnecessary Crisis—Setting the Record Straight About Iran's Nuclear Program," in the New York Times, November 18, 2005. P. A11.

[17] Stern, Roger. "The Iranian Petroleum Crisis and United States National Security," *Proceedings of the National Academy of Sciences of the United States of America*. December 26, 2006.

[18] "Leader Says West Knows Iran Not Seeking 'Nuclear Weapons.'" Vision of the Islamic Republic of Iran Network, February 22, 2012.

[19] For information on Stuxnet and its origins and effects, see Broad William, John Markoff and David Sanger. "Israeli Test on Worm Called Crucial in Iran Nuclear Delay." *New York Times*, January 15, 2011.

[20] Joby Warrick. "Iran Targets Nuclear Agency." Washington Post, October 8, 2012.

[21] http://isis-online.org/uploads/isis-reports/documents/Iran_report_—_August_30_2012.pdf

[22] IAEA report of February 25, 2011. http://lewis.armscontrolwonk.com/files/2011/02/gov2011-7.pdf.

[23] For text of the agreement, see http://www.iaea.org/NewsCenter/Focus/IaeaIran/eu_iran 14112004.shtml. EU-3-Iran negotiations on a permanent nuclear pact began on December 13, 2004, and related talks on a trade and cooperation accord (TCA) began in January 2005.

[24] In November 2006, the IAEA, at U.S. urging, declined to provide technical assistance to the Arak facility.

[25] Voting in favor: United States, Australia, Britain, France, Germany, Canada, Argentina, Belgium, Ghana, Ecuador, Hungary, Italy, Netherlands, Poland, Portugal, Sweden, Slovakia, Japan, Peru, Singapore, South Korea, India. Against: Venezuela. Abstaining: Pakistan, Algeria, Yemen, Brazil, China, Mexico, Nigeria, Russia, South Africa, Sri Lanka, Tunisia, and Vietnam.

[26] Voting no: Cuba, Syria, Venezuela. Abstaining: Algeria, Belarus, Indonesia, Libya, South Africa.

[27] See http://daccessdds.un.org/doc/UNDOC/GEN/N06/290/88/PDF/N0629088.pdf? Open Element.

[28] One source purports to have obtained the contents of the package from ABC News: http://www.basicint.org/pubs/Notes/BN060609.htm.

[29] Dempsey, Judy. "U.S. Urged to Talk With Iran." *International Herald Tribune*, February 5, 2009.

[30] Text of the pact is at http://www.cfr.org/publication/22140/.

[31] It was adopted by a vote of 12-2 (Turkey and Brazil voting no) with one abstention (Lebanon)

[32] Text of the resolution is at http://www.isis-online.org/uploads/isis-reports/documents/Draft_ resolution_on_Iran_ annexes.pdf.

[33] http://vienna.usmission.gov/110309p51.html.

[34] Deputy Director of National Intelligence for Analysis. "Unclassified Report to Congress on the Acquisition of Technology Relating to Weapons of Mass Destruction and Advanced Conventional Munitions, Covering 1 January to 31 December 2010. March 2011.

[35] http://thehill.com/blogs/defcon-hill/policy-and-strategy/207275-panetta-iran-could-havenuclear-weapons-delivery- vehicles-in-2-3-years.

[36] Annual Threat Assessment of the Intelligence for the Senate Select Committee on Intelligence, Dennis C. Blair, Director of National Intelligence, February 2, 2010.

[37] Broad, William and David Sanger. "Relying On Computer, U.S. Seeks to Prove Iran's Nuclear Aims." *New York Times*, November 13, 2005.

[38] Walsh, Elsa. "Annals of Politics: Louis Freeh's Last Case." *The New Yorker*, May 14, 2001. The June 21, 2001, federal grand jury indictments of 14 suspects (13 Saudis and a Lebanese citizen) in the Khobar bombing indicate that Iranian agents may have been involved, but no indictments of any Iranians were announced. In June 2002, Saudi Arabia reportedly sentenced some of the eleven Saudi suspects held there. The 9/11 Commission final report asserts that Al Qaeda might have had some as yet undetermined involvement in the Khobar Towers attacks.

[39] "Iran Spy Cell Dismantled in Kuwait." Associated Press, May 6, 2010; "Iran Cell Planned Attacks in Kuwait, Minister Says. *Reuters*, April 21, 2011.

[40] A rebellion in northern Yemen was begun by dissident Shiite cleric Hussein Badruddin Al Houthi, and the rebellion is widely referred to as the Houthi rebellion. Eric Schmitt and Robert F. Worth, "With Arms for Yemen Rebels, Iran Seeks Wider Mideast Role," *New York Times*, March 15, 2012.

[41] Background on this issue is covered in CRS Report RS22323, *Iran-Iraq Relations*, by Kenneth Katzman.

[42] Michael Gordon. "Iran Supplying Syrian Military Via Iraqi Airspace." *New York Times*, September 5, 2012.

[43] CNN "Late Edition" interview with Hamas co-founder Mahmoud Zahar, January 29, 2006.

[44] Hamas Leader Praises Iran's Help in Gaza 'Victory.' CNN.com, February 1, 2009.

[43] CNN "Late Edition" interview with Hamas co-founder Mahmoud Zahar, January 29, 2006.

[44] Hamas Leader Praises Iran's Help in Gaza 'Victory.' *CNN.com*, February 1, 2009.

[45] For detail on Hezbollah, see CRS Report R41446, *Hezbollah: Background and Issues for Congress*, by Casey L. Addis and Christopher M. Blanchard.

[46] "Israel's Peres Says Iran Arming Hizbollah." *Reuters*, February 4, 2002.

[47] Neil MacFarquhar. "Iran is Seeking Lebanon Stake As Syria Totters." *New York Times*, May 25, 2012.

[48] Rotella, Sebastian. "In Lebanon, Hezbollah Arms Stockpile Bigger, Deadlier." *Los Angeles Times*, May 4, 2008.

[49] Shadid, Anthony. "Armed With Iran's Millions, Fighters Turn to Rebuilding." *Washington Post*, August 16, 2006.

[50] Adam Entous and Matthew Rosenberg. "U.S. Says Iran Helps Crackdown in Syria." *Wall Street Journal*, April 14, 2011.

[51] Hugh Tomlinson. "Tehran Split Over Billions Spent by Spy Chief to Prop Up Assad Regime." London Times, October 1, 2012.

[52] Levinson, Charles. "Iran Arms Syria With Radar." *Wall Street Journal*, June 30, 2010.

[53] James Ball. "Documents: Iran Helped Syria Expand Chemical Weaponry." *Washington Post*, July 28, 2012.

[54] Substantially more detail on Iran's activities in Afghanistan is contained in: CRS Report RL30588, *Afghanistan: Post-Taliban Governance, Security, and U.S. Policy*, by Kenneth Katzman.

[55] Matthew Rosenberg and Annie Lowry. "Iranian Currency Traders Find a Haven in Afghanistan" New York Times, August 18, 2012.

[56] Maria Abi-Habib. "Tehran Builds On Outreach to Taliban." *Wall Street Journal*, August 1, 2012.

[57] Lancaster, John and Kamran Khan."Pakistanis Say Nuclear Scientists Aided Iran." *Washington Post*, January 24, 2004. http://isis-online.org/uploads/isis-reports/documents/IAEA_Iran_ 8Nov2011.pdf.

[58] Gertz, Bill. "Al Qaeda Terrorists Being Held by Iran." *Washington Times*, July 24, 2003.

[59] Gertz, Bill. "CIA Points to Continuing Iran Tie to Al Qaeda." *Washington Times*, July 23, 2004.

[60] "Tehran Pledges to Crack Down on Militants." Associated Press, July 18, 2005.

[61] Martin Arostegui. "Iran Tries to Gain Sway in Latin America." Wall Street Journal, December 6, 2011.

[62] An exception was the abortive 1985-1986 clandestine arms supply relationship with Iran in exchange for some American hostages held by Hezbollah in Lebanon (the so-called "Iran-Contra Affair"). Iran has an interest section in Washington, DC, under the auspices of the Embassy of Pakistan; it is staffed by Iranian Americans. The U.S. interest section in Tehran has no American personnel; it is under the Embassy of Switzerland.

[63] Sciolino, Elaine. *The Outlaw State: Saddam Hussein's Quest for Power and the Gulf Crisis.* New York: John Wiley and Sons, 1991. p. 168.

[64] Cooper, Helene and David Sanger. "Strategy on Iran Stirs New Debate at White House." *New York Times*, June 16, 2007.

[65] These were prominent themes in speeches by President Bush such as at the Merchant Marine Academy on June 19, 2006, and his September 18, 2006, speech to the U.N. General Assembly.

[66] Wright, Robin. "U.S. In 'Useful' Talks With Iran." *Los Angeles Times*, May 13, 2003.

[67] http://www.armscontrol.org/pdf/2003_Spring_Iran_Proposal.pdf.

[68] Helene Cooper and Mark Landler. "U.S. Officials Say Iran Has Agreed to Nuclear Talks." New York Times, October 21, 2012.

[69] Jeffrey Goldberg. "Obama to Iran and Israel: 'As President of the United States, I Don't Bluff'" *The Atlantic*, March 2, 2012.

[70] http://2scottmontgomery.blogspot.com/2011/12/panetta-brookings-speech.html.

[71] Joby Warrick. "Iran: Underground Sites Vulnerable, Experts Say." *Washington Post*, March 1, 2012. For an extended discussion of U.S. air strike options on Iran, see Rogers, Paul. *Iran: Consequences Of a War*. Oxford Research Group, February 2006.

[72] See also, Washington Institute for Near East Policy. "The Last Resort: Consequences of Preventive Military Action Against Iran," by Patrick Clawson and Michael Eisenstadt. June 2008.

[73] For more information on this and other U.S. sales to the UAE, see CRS Report RS21852, *The United Arab Emirates (UAE): Issues for U.S. Policy*, by Kenneth Katzman.

[74] David Sanger and Eric Schmitt. "To Calm Israel, U.S. Offers Ways to Restrain Iran." New York Times, September 3, 2012.

[75] Jay Solomon and Julian Barnes. "U.S. Weighs A Direct Line to Tehran." *Wall Street Journal*, September 19, 2011.

[76] This option is analyzed in substantial depth in: CRS Report R42443, *Israel: Possible Military Strike Against Iran's Nuclear Facilities*, coordinated by Jim Zanotti.

[77] Jeffrey Goldberg. "Obama to Iran and Israel…" op. cit.

[78] Sanger and Schmitt, New York Times, September 3, 2012, op. cit.

[79] See http://www.theatlantic.com/magazine/archive/2010/09/the-point-of-no-return/8186/

[80] CRS conversations with U.S. officials responsible for Iran policy. 1980-1990. After a period of suspension of such assistance, in 1995, the Clinton Administration accepted a House-Senate conference agreement to include $18-$20 million in funding authority for covert operations against Iran in the FY1996 Intelligence Authorization Act (H.R. 1655, P.L. 104-93), according to a *Washington Post* report of December 22, 1995. The Clinton Administration reportedly focused the covert aid on changing the regime's behavior, rather than its overthrow.

[81] White House, Office of the Press Secretary. "Statement by the President on the Attempted Attack on Christmas Day and Recent Violence in Iran." December 28, 2009.

[82] http://www.whitehouse.gov/the-press-office/2011/02/27/statement-national-security-council-spokesman-tommy-vietor-iran.

[83] White House. "Remarks of President Obama Marking Nowruz." March 20, 2011.

[84] This legislation was a modification of H.R. 282, which passed the House on April 26, 2006, by a vote of 397-21, and S. 333, which was introduced in the Senate.

[85] Ross, Brian and Richard Esposito. Bush Authorizes New Covert Action Against Iran, http://blogs.abcnews.com/theblotter/2007/05/bush_authorizes.html.

[86] Three other Iranian Americans were arrested and accused by the Intelligence Ministry of actions contrary to national security in May 2007: U.S. funded broadcast (Radio Farda) journalist Parnaz Azima (who was not in jail but was not allowed to leave Iran); Kian Tajbacksh of the Open Society Institute funded by George Soros; and businessman and peace activist Ali Shakeri. Several congressional resolutions called on Iran to release Esfandiari (S.Res. 214 agreed to by the Senate on May 24; H.Res. 430, passed by the House on June 5; and S.Res. 199). All were released by October 2007. Tajbacksh was rearrested in September 2009 and remains incarcerated.

[87] CRS conversation with U.S. officials of the "Iran Office" of the U.S. Consulate in Dubai. October 2009.

[88] Solomon, Jay. "U.S. Shifts Its Strategy Toward Iran's Dissidents." *Wall Street Journal*, June 11, 2010.

[89] For more discussion of such legislation, see CRS Report RS20871, *Iran Sanctions*.

[90] Ibid.

[91] The conference report on the FY2006 regular foreign aid appropriations, P.L. 109-102, stated the sense of Congress that such support should be considered.

[92] CRS conversations with Iranian members of the Green Movement. December 2009-August 2010.

[93] http://ipsnews.net/news.asp?idnews=54504. Confirmed to CRS on February 25, 2011, by a member of the BBG.

[94] Stockman, Farah. "'Long Struggle' With Iran Seen Ahead." *Boston Globe*, March 9, 2006.

INDEX

#

20th century, 64
9/11, 116, 141
9/11 Commission, 116, 141

A

accelerometers, 41
access, viii, 2, 28, 38, 41, 73, 115
accommodations, 126
accounting, 13, 53, 77
activity level, 83
adaptation, 39
adaptations, 39
advancement, 9, 134
advancements, 9
aerospace, 52
Afghanistan, ix, 3, 46, 63, 64, 76, 88, 90, 103, 111, 113, 114, 115, 116, 121, 124, 127, 129, 142
Africa, 116, 119
agencies, 4, 11
aggression, 45
Air Force, 13, 16, 17, 26, 44, 52, 53, 72, 88, 89, 104, 128
Al Qaeda, 104, 116, 119, 141, 142
Alaska, 51, 52
Algeria, 141
ambassadors, 114
ammonium, 41
anti-drug efforts, 137
apex, 66
appointments, 68, 69
appraisals, 36
appropriations, 3, 131, 143
Arab world, 77
architect, 73, 111
Argentina, 117, 118, 141
armed forces, 13, 66, 87, 89
armed groups, 3, 82, 130
Armenia, 112
Armenians, 112
arms control, 10, 46
arms sales, 40, 94, 119, 120, 126
arms transfer agreements, 36, 39
arrest, 72, 73, 74, 78, 84, 131
arrests, 66, 78, 84, 86
Asia, 33, 58, 73
assassination, 67, 76, 105
assessment, 12, 20, 21, 24, 27, 28, 29, 46, 51, 52, 55, 66, 71, 87, 88, 104, 127, 128
assets, 5, 91, 95, 101, 106, 112, 130, 137
asylum, 81
atmosphere, 32, 34, 51, 85
atrocities, 115
attitudes, 42
Austria, 96
authorities, 127
authority, 66, 67, 68, 71, 76, 81, 90, 97, 106, 109, 127, 133, 136, 141, 143
autonomy, 86
aversion, 122
Azerbaijan, 88, 105, 112, 113, 133

B

Bahrain, ix, 46, 63, 64, 103, 105, 107, 123, 125, 126
bail, 86
ballistic missiles, vii, 1, 2, 4, 5, 6, 7, 11, 12, 13, 15, 17, 18, 20, 22, 24, 25, 28, 29, 34, 35, 36,

38, 41, 42, 45, 46, 47, 48, 52, 56, 59, 61, 101, 102, 124, 127
Baluchistan, 30, 31, 64, 82, 115
ban, 86, 94, 95, 97, 101, 120, 121, 130, 135, 136, 137
banking, 98, 106, 108
banks, 91, 97, 101, 136
base, vii, 1, 7, 14, 22, 46, 68, 70, 82, 91, 103, 106, 114, 118
Beijing, 60, 113
Belarus, 89, 141
Belgium, 141
benefits, 123, 138
bilateral talks, ix, 64, 98, 123
biological weapons, 47
blogger, 74, 85
blogs, 22, 85, 141, 143
Bolivia, 117, 118
border security, 113
Brazil, 97, 117, 118, 138, 141
breakdown, 13, 54
bribes, 114
Britain, ix, 24, 56, 64, 72, 93, 94, 98, 114, 125, 141
browsing, 132
Bulgaria, 105
burn, 51
businesses, 53, 83, 85, 90

C

Cabinet, 69
cables, 21
Cairo, 122
candidates, ix, 64, 66, 67, 69, 70, 75, 77, 78, 140
capsule, 34
carbon, 41
carbon materials, 41
cargoes, 101
cash, 35, 113
Caspian Sea, 112
casting, 117
Caucasus, 64
ceasefire, 81, 83
censorship, 131
Central Asia, 90, 112, 113, 115
certification, 137
challenges, 6, 7, 21, 27, 42, 87, 112
checks and balances, 66
chemical, 47, 48, 101, 112, 129
Chicago, 5, 51, 115
Chief of Staff, 57
children, 71, 72, 82
Chile, 117
China, viii, 2, 8, 9, 21, 24, 25, 33, 37, 38, 39, 40, 41, 49, 52, 57, 59, 94, 95, 97, 113, 138, 141
Chinese government, 39
Christians, 69, 85
CIA, 12, 53, 65, 138, 142
CIS, 57, 73
cities, 55, 71, 78, 89, 110
citizens, 110, 132, 137
City, 104
civil servants, 107
civil society, 131, 134
civil war, 15, 41
classes, 73, 76
clients, 15, 74
Clinton Administration, 120, 126, 137, 143
closure, 98
clothing, 73
CNN, 84, 133, 142
Cold War, 5, 42, 58, 59
collaboration, 33, 38, 39
Colombia, 117
commander-in-chief, 66, 67
commercial, 57, 103
Committees on Appropriations, 47
communication, 78, 121, 131
communications channel, 127
communities, 43, 85
community, ix, 3, 6, 11, 12, 27, 28, 36, 38, 44, 63, 67, 85, 98, 100, 106, 110, 119, 122, 128
competition, 113
complexity, 8, 52
compliance, 80, 94, 95, 96, 99, 101, 111
computer, 92, 102, 132
conference, 76, 114, 123, 134, 143
confessions, 110
conflict, 4, 68, 105, 108, 109, 110, 111, 115, 125, 127
confrontation, ix, 64, 67, 87, 127
Congress, vii, ix, 1, 2, 3, 5, 12, 14, 23, 24, 27, 36, 37, 38, 40, 42, 47, 48, 52, 53, 54, 56, 57, 58, 59, 60, 63, 64, 117, 119, 120, 125, 126, 127, 130, 141, 142, 143
consensus, 25, 28, 42, 67, 73, 93, 117
consent, 69
construction, 22, 24, 55, 90, 113
contingency, 127

control measures, viii
controversial, 20, 69, 76, 110
conversations, 138, 143
cooperation, 4, 7, 11, 30, 35, 36, 37, 38, 39, 42, 51, 82, 87, 91, 92, 95, 106, 113, 115, 141
coordination, 128
corruption, 70
cost, 27, 90, 92
Court of Appeals, 80
covering, 43, 44, 132
credentials, 86
crimes, 86
criticism, 67, 68, 76, 80, 84, 99, 109, 110, 111, 123, 131, 132
crowds, 111
crude oil, 118
cruise missiles, 12, 47, 48, 88, 101, 102, 109
CSS, 16, 49, 52, 102
Cuba, 90, 91, 117, 118, 141
cultural influence, 67
culture, 72, 86, 138
currency, ix, 64, 75, 77, 111, 113
customers, 36, 41, 105, 135, 138
Cyprus, 105, 109
Czech Republic, 127

D

damages, 116
database, 53
deaths, 71, 81, 86
debt servicing, 58
decision makers, 3, 7, 19, 42
defence, 53, 54, 60
deficiencies, 119, 133
democracy, 123, 130, 131, 132, 133, 134
Democratic Republic of Congo, 119
demonstrations, 65, 71, 73, 78, 105, 129, 130
denial, 29, 136
Department of Defense, 32, 45, 51, 52, 57, 61, 141
deployments, 7, 128
depth, 143
derivatives, 39
destruction, 42
detainees, 81
detection, 88
detention, 84
deterrence, 5, 27, 42, 43, 51
detonation, 89
developing countries, 40
diplomacy, 27, 93, 106, 121, 122, 124, 131, 133, 134
directors, 90
discrimination, 84, 86
dissatisfaction, 65
dissidents, 65, 72, 74, 84, 105, 107, 108, 118, 129
distortions, 11
diversity, 65, 67
donors, 138
draft, 25, 84, 96, 97, 127
drawing, 119
dual-use items, 48

E

economic integration, 105
economic problem, 111
Ecuador, 90, 91, 117, 118, 141
editors, 85
education, 90
Egypt, 36, 78, 91, 103, 109, 129
elders, 73
election, ix, 64, 65, 66, 67, 68, 69, 70, 71, 72, 73, 74, 76, 77, 78, 80, 84, 86, 94, 96, 117, 120, 122, 123, 130, 131, 135, 140
electoral college, 69
electricity, 89, 91, 92, 110
embargo, ix, 63, 64, 98, 99, 100, 119, 123, 125
embassy, 77, 98, 104, 116, 117, 118, 135
employees, 132, 133
employment, 133
endangered, 3
enemies, 112
energy, ix, 18, 64, 87, 89, 91, 93, 94, 95, 107, 120, 136
enforcement, viii, 2, 39, 40, 85, 101, 106
engineering, 7, 8, 32, 39, 97
England, 51, 57
environment, 131
environmental impact, 3
equipment, 19, 29, 32, 33, 37, 40, 41, 60, 95, 111, 113, 118, 126, 130, 131, 136, 137
espionage, 86
estrangement, 103, 119, 120, 122
ethnicity, 113
EU, 80, 93, 94, 95, 97, 98, 99, 100, 115, 120, 123, 125, 135, 136, 141
Europe, viii, 2, 4, 5, 12, 20, 27, 29, 51, 56, 57, 61, 73, 74, 80, 86, 90, 102, 105, 119, 124, 131, 132

European Union, ix, 63, 64, 80, 112
evidence, 21, 25, 29, 32, 35, 39, 42, 56, 80, 83, 93, 103, 106, 107, 118, 128, 130, 131
evil, 121
evolution, 130
exaggeration, 60
executive branch, 67
Executive Order, 80, 83, 91, 112, 114, 116, 130, 132, 135, 136, 137
executive orders, 137
exercise, 43, 44, 45, 51
exile, ix, 64, 72, 74, 77, 78, 79, 132, 133
expenditures, 12, 53, 68
expertise, 7, 37, 39, 48, 58, 116
exploitation, 86
explosives, 114, 118
export control, viii, 2, 7, 37, 39, 40, 59
exports, ix, 4, 40, 41, 63, 65, 88, 91, 106, 109, 119, 132, 136, 138
expulsion, 81, 106

F

Facebook, 70, 71
factories, 74
fairness, 75
faith, 75, 85
family members, 72, 116
FBI, 86
fear, 4, 67, 73, 74, 85, 115, 138
fears, 75, 81, 92, 106, 111, 112
feminism, 79
financial, ix, 8, 64, 68, 95, 101, 107, 110, 113, 136
financial system, 136
fires, 53, 54, 56, 60, 61
first generation, 20
flexibility, 81, 98
flight, viii, 2, 5, 7, 12, 15, 17, 18, 19, 20, 26, 33, 34, 37, 51, 52, 102, 124
flights, 25, 34, 108
food, 136
force, viii, 2, 4, 6, 7, 13, 35, 64, 71, 77, 81, 88, 103, 110, 112, 125, 127
foreign affairs, 54
foreign aid, 134, 143
foreign assistance, viii, 2, 5, 8, 27, 136
foreign banks, 136
Foreign Broadcast Information Service, 58, 59
foreign companies, 9
foreign firms, 137
foreign investment, 120, 136
foreign language, 133
foreign person, 48, 137
foreign policy, 67, 77, 87, 90, 94, 98, 99, 103, 104
formation, 110
formula, 110
foundations, 68, 90
France, 8, 9, 10, 25, 53, 57, 60, 65, 79, 80, 93, 94, 96, 105, 138, 141
fraud, 70, 71
freedom, 85, 86, 132, 134
Freedom Support Act, 130
fruits, 136
funding, 3, 9, 37, 38, 109, 111, 129, 130, 131, 143
funds, 68, 114, 116, 130, 131, 134

G

Gaza Strip, 103, 109
GDP, 12, 53, 88, 138
Geneva Convention, 81
genocide, 76
geometry, 19
Georgia, 96, 105
Germany, 55, 57, 59, 64, 83, 93, 94, 96, 98, 133, 138, 141
global trade, ix, 64
God, 57, 73, 76
goods and services, 36
governance, 133
government budget, 68
governments, 9, 28, 46, 57, 78, 93
governor, 76
GPS, 33
grand jury, 141
grants, 95
grouping, 42, 74, 113
growth, 5, 12, 17, 52
guidance, 8, 17, 19, 29, 41, 72
guidelines, 39, 59

H

Hamas, 3, 103, 109, 119, 142
hard currency, 37, 58, 106, 111
hardliners, ix, 3, 64, 67, 69, 71, 72, 73, 76, 90, 96, 116, 123
health, 131
health care, 131

heat shield, 60
heavy artillery rockets, vii, 1
hegemony, 51
Hezbollah, 3, 15, 54, 90, 95, 103, 104, 105, 108, 109, 110, 111, 120, 142
history, 9, 86
homelessness, 133
homes, 111
homosexuality, 76
host, 91, 112
hostilities, 109, 125
hotels, 138
House, 3, 28, 46, 47, 48, 53, 56, 58, 76, 80, 85, 117, 125, 127, 140, 143
housing, 104, 116
hub, 119
human, 72, 74, 80, 83, 84, 85, 86, 87, 91, 112, 120, 123, 129, 130, 131, 133, 134, 137
human resources, 133
human right, 72, 74, 80, 83, 84, 85, 86, 87, 91, 112, 120, 123, 129, 130, 131, 134, 137
human rights, 72, 74, 80, 83, 84, 85, 86, 87, 91, 112, 120, 123, 129, 130, 131, 134, 137
Hungary, 141
husband, 86
hypocrisy, 129

I

ideal, 89
ideals, 90
identification, 111
ideology, 71, 91, 103
image, 76
imagery, 14, 22, 24, 33, 55
IMF, 138
immigration, 86
immunity, 128
impeachment, 69
imports, 135, 136
imprisonment, 66
improvements, 12, 101
inauguration, 77, 107, 120
incumbents, 69
independence, 107
Independence, 122
India, 9, 10, 52, 53, 56, 57, 89, 105, 113, 115, 138, 141
individuals, 36, 38, 95, 134
Indonesia, 90, 95, 141
industries, 8, 37, 40
industry, 35, 55, 65, 73, 106
infrastructure, 9, 11, 12, 15, 22, 34, 53, 118, 124
inheritance, 85
injure, 128
inspections, 94, 95, 97, 99, 100
institutions, 66, 68, 90, 108
intelligence, viii, 2, 6, 9, 11, 12, 19, 21, 27, 28, 29, 36, 40, 43, 44, 53, 54, 55, 65, 68, 74, 76, 90, 92, 93, 102, 104, 106, 116, 120, 127
intelligence gathering, viii, 2
intercontinental ballistic missile (ICBM), viii, 2, 5, 27
interference, 4, 87, 92, 130
international affairs, 11
International Atomic Energy Agency, 89
International Covenant on Civil and Political Rights, 86
international diplomacy, 121
international financial institutions, 95, 136
International Financial Institutions, 135
international meetings, 122
international pressure, ix, 63, 75
international standards, 84
international terrorism, 47, 79, 104, 110, 114, 137
interpersonal relations, 115
intervention, 27, 76, 105, 112
investment, 11, 51, 120, 136, 137
Iran Sanctions Act, 115, 118, 120, 136
Iraq, viii, 3, 8, 15, 36, 38, 47, 54, 57, 58, 63, 65, 70, 73, 81, 83, 86, 88, 90, 95, 99, 103, 104, 107, 108, 113, 116, 120, 121, 124, 127, 128, 137, 142
Iraq War, 15, 57, 65, 70, 73, 107, 120
irony, 41
Islam, 54, 64, 67, 71, 76, 84, 91, 140, 141
Islamic law, 66
Islamic state, 72
Islamic world, 11
Islamism, 79
islands, 106
isolation, 72, 73, 119, 124
Israel, vii, viii, 2, 4, 10, 12, 15, 17, 20, 22, 29, 32, 45, 46, 54, 55, 57, 60, 61, 63, 64, 67, 73, 76, 85, 86, 87, 102, 103, 105, 109, 110, 111, 113, 119, 122, 123, 124, 126, 127, 128, 129, 142, 143
issues, 6, 27, 67, 68, 72, 73, 78, 80, 83, 84, 100, 115, 118, 121, 122, 123, 131, 133, 138
Italy, 33, 138, 141

J

Japan, 57, 98, 138, 141
Jews, 69, 85, 86
joint ventures, 118
Jordan, 103
journalists, 78, 85, 99, 111, 131
judiciary, 66, 72, 75, 84, 86, 140
justification, 112

K

Kazakhstan, 113
Kenya, 105, 116
Korea, 19, 21, 29, 35, 36, 38, 52, 55, 89, 102, 138
Kurds, 83, 85, 86
Kuwait, 46, 104, 105, 107, 126, 141
Kyrgyzstan, 113

L

Latin America, 117, 118, 142
law enforcement, 88
laws, viii, 2, 40, 136, 137
lawyers, 140
lead, 81, 98, 106, 114, 121
leadership, ix, 11, 30, 63, 72, 73, 77, 90, 99, 112, 116, 125, 133, 135
leadership development, 90
Lebanon, 54, 90, 103, 104, 105, 110, 120, 141, 142
legislation, ix, 40, 58, 64, 66, 69, 127, 130, 131, 143
lending, 95, 101, 135
letters of credit, 75
light, 31, 94, 96, 97, 101, 109, 121, 140
loans, 95, 136
logistics, 15, 43
loyalty, 67, 90

M

majority, vii, ix, 1, 46, 63, 69, 75, 83
man, 66, 76
management, 8, 9, 90
manipulation, 64
manufacturing, 8, 19
maritime security, 126
marketing, 35
marketplace, 37, 58
marriage, 71
masking, 34
mass, 15, 29, 131
materials, viii, 2, 12, 28, 29, 34, 35, 36, 37, 40, 41, 47, 75
matter, vii, 2, 32, 41, 44, 125, 137
media, 11, 15, 22, 24, 25, 43, 44, 53, 55, 61, 70, 78, 86, 90, 99, 111, 123, 130, 134, 141
medical, 89, 91, 93, 96, 99, 100, 120, 136
medium-range ballistic missile (MRBM), viii, 2
MEK, 140
membership, 98, 113
mentor, 72
Mexico, 117, 141
Middle East, vii, viii, 1, 2, 3, 4, 12, 17, 35, 39, 43, 46, 53, 54, 59, 60, 61, 63, 85, 88, 103, 119, 120, 123, 126, 129
military aid, 107, 110
military exercises, 20, 21, 42, 43, 44, 45, 46, 102
military-to-military, 89, 115
militia, 89, 98, 119
militias, 95, 108, 124
miniaturization, 6, 52
minorities, 65, 69, 82
minors, 86
missile defenses, 4
mission, 19, 34, 57, 76, 83, 84, 87, 125, 132
missions, 7, 88, 133
models, 8, 15, 20
moderates, 103
modernization, 6
modifications, 19, 32
money laundering, 136
monopoly, 85
Moon, 103
Moscow, 25, 99, 100
murder, 86
Muslim states, 85
Muslims, 82, 85, 103, 105, 107

N

naming, 20, 40, 106
narcotics, 113, 137
National Academy of Sciences, 91, 141
National Defense Authorization Act, 47, 48, 98, 125, 127, 141
National Intelligence Estimate, 38, 56
national interests, 103

national security, vii, 3, 4, 6, 27, 66, 87, 138, 143
National Security Council, 66, 72, 129
nationalists, 64, 112
NATO, 5, 51, 114
natural gas, 106, 135
negotiating, 96, 98, 123, 128, 135
negotiation, 95
Netherlands, 141
next generation, 27
Nicaragua, 117, 118
Nigeria, 98, 116, 119, 141
Nobel Prize, 74
North Africa, 88
North Atlantic Treaty Organization, 5
North Korea, 3, 15, 17, 19, 20, 21, 27, 28, 29, 35, 36, 38, 40, 48, 52, 54, 55, 56, 58, 88, 89, 102, 112, 121, 137
NPT, 93, 94, 99, 122
nuclear arms race, 87
nuclear program, vii, ix, 2, 3, 6, 42, 52, 63, 86, 89, 90, 91, 92, 93, 98, 99, 105, 113, 119, 121, 122, 124, 127, 128, 138
nuclear talks, ix, 63, 94, 98, 121, 123, 135
nuclear weapon, vii, viii, 1, 4, 6, 8, 19, 24, 25, 27, 31, 42, 48, 54, 56, 63, 67, 87, 89, 91, 92, 93, 100, 105, 115, 122, 124, 125, 126, 128
nuclear weapons, vii, 1, 4, 6, 8, 19, 24, 25, 27, 31, 42, 48, 54, 56, 89, 91, 92, 93, 100, 115, 124, 125, 126

O

Obama, viii, 3, 63, 81, 87, 96, 97, 118, 122, 123, 124, 126, 127, 128, 129, 131, 135, 142, 143
Obama Administration, viii, 3, 63, 81, 87, 96, 97, 118, 122, 126, 127, 128, 129, 131, 135
oil, ix, 4, 63, 65, 87, 88, 90, 91, 98, 99, 100, 105, 106, 107, 113, 115, 117, 120, 123, 124, 125, 135, 136, 138
oil production, 4, 105
omission, 15
openness, 10
Operation Iraqi Freedom, 81
operations, viii, 2, 43, 45, 68, 83, 93, 113, 129, 130, 134, 143
opportunities, 68, 133
opposition movement, ix, 63, 65, 71, 80
oppression, 74, 86
optimism, 99
orbit, viii, 2, 11, 28, 31, 32, 51, 53, 102
organize, 67, 110, 112
organs, 67
outreach, 119, 135
overlay, 108
oversight, 42, 66, 68, 84

P

Pacific, 33, 58
Pakistan, 8, 30, 40, 41, 52, 54, 82, 86, 113, 115, 116, 141, 142
Palestinian Islamic Jihad, 109
Parliament, 67
Pashtun, 113, 114
peace, 51, 103, 108, 120, 138, 143
peace process, 103, 120
pedigree, 20
penalties, 91
Pentagon, 4, 5, 7, 29, 43, 51, 53, 57, 60, 102
permission, 68, 75, 108, 122
Persian Gulf, vii, 1, 4, 5, 30, 44, 46, 53, 61, 64, 79, 86, 87, 89, 90, 92, 94, 100, 105, 106, 120, 124, 126, 127, 138
personal communication, 132
Peru, 141
petroleum, 136
Petroleum, 112, 141
photographs, 44
physics, 6
pipeline, 106, 113, 115
plants, 10
platform, 10, 126
plutonium, 89
PMOI, 65, 79, 80, 81, 82, 89, 105
Poland, 51, 126, 141
police, 72, 78, 81
policy, viii, 3, 24, 25, 30, 42, 63, 68, 103, 104, 109, 112, 114, 119, 120, 121, 122, 125, 128, 129, 130, 132, 133, 141, 143
policy making, 30
policy options, 120
policymakers, 3
political opposition, 73, 96
political parties, 68
political party, 134
political pluralism, 74
political system, 65, 90
politics, 72, 110
polling, 70, 71
Popular Front, 109

population, 66, 69, 85, 86, 103, 112, 133, 134
Portugal, 141
potential benefits, 123
prayer, 71
PRC, 39, 40
predictability, 8
preparation, 17, 33, 34, 99, 100
presidency, 68, 75, 94, 106
president, 66, 67, 68, 69, 71, 72, 73, 76, 97, 110, 117
President, ix, 3, 30, 33, 34, 46, 47, 48, 57, 64, 65, 67, 68, 69, 72, 79, 82, 88, 96, 98, 103, 105, 108, 109, 110, 112, 113, 114, 117, 118, 119, 120, 121, 122, 123, 124, 125, 127, 128, 129, 130, 131, 132, 133, 136, 142, 143
President Clinton, 121
President Obama, 67, 96, 98, 117, 122, 123, 124, 125, 128, 129, 130, 131, 132, 143
presidential authority, 68
prestige, 27
principles, 7, 76
prisoners, 104
private sector, 30
probability, 27
probe, 102
procurement, 90
producers, 119
profit, 40
programming, 132, 134
project, 8, 101, 110, 115
proliferation, 5, 6, 8, 9, 10, 17, 37, 38, 39, 40, 41, 42, 47, 48, 51, 91, 95, 137, 141
protection, 82
prototypes, 29
public domain, 6
public policy, 38
public support, 130
publishing, 75
punishment, 109
purchasing power, 138
purchasing power parity, 138
purity, ix, 63, 66

Q

quality production, 41
questioning, 75, 76

R

race, 73
radar, 11, 43, 53, 88, 112, 126
radicals, 65
radio, 79, 83, 132
radius, 15
Ramadan, 100
reading, 132
reality, 28, 118
reciprocity, 99
recognition, 104
recommendations, 101
reconciliation, 73, 114
recovery, 18
reelection, 72, 107
reform, 73, 74, 77, 85
Reform, 53
reforms, 112
refugee status, 81, 82
regional policy, 109
Registry, 88, 97
relatives, 116, 133
relaxation, 74
reliability, 7, 10, 42
relief, ix, 63, 113
religion, 76, 85
reporters, 15
repression, 65, 74, 78, 85
reprocessing, 95
Republican Party, 65
reputation, viii, 2, 21
requirements, 3, 47, 67, 131
resettlement, 82
resistance, 80, 81, 125
resolution, 33, 54, 55, 56, 57, 59, 64, 87, 90, 91, 97, 134, 141
resources, 9, 28, 34, 90, 91, 111, 112
response, 3, 29, 45, 46, 58, 61, 93, 94, 98, 100, 107, 112, 120, 125, 131, 135
restrictions, 73, 122, 123, 130, 132, 137
retaliation, 79, 88, 104, 105, 108, 124, 125
Revolutionary Guard, 13, 54, 66, 67, 68, 70, 72, 74, 76, 82, 87, 88, 89, 90, 91, 97, 101, 136, 141
rhetoric, 85
rights, 83, 84, 85, 86, 87, 112, 123, 129, 130, 131, 134
risk, 77, 104
risks, 128
rods, 93

Romania, 127
routes, 115
rubber, 69
rule of law, 133
rules, 74
runoff, 70, 75
rural areas, 71, 75, 77
Russia, viii, 2, 4, 7, 8, 9, 21, 24, 25, 28, 29, 33, 36, 37, 38, 39, 42, 51, 57, 64, 88, 89, 93, 94, 95, 96, 97, 98, 113, 127, 138, 141

S

sabotage, 92, 103
Saddam Hussein, 80, 107, 108, 142
safety, 31, 82, 99
sanctions, vii, ix, 1, 2, 3, 7, 25, 27, 33, 39, 40, 41, 42, 47, 53, 63, 64, 67, 74, 83, 85, 88, 90, 92, 94, 95, 96, 97, 98, 99, 100, 101, 102, 103, 105, 106, 108, 111, 113, 115, 117, 118, 119, 120, 121, 122, 123, 124, 125, 127, 128, 130, 135, 136, 137
satellite service, viii, 2
Saudi Arabia, 46, 52, 104, 105, 114, 116, 117, 126, 141
school, 38, 39, 72, 86
science, 7, 40, 89, 131
SCO, 113
second generation, 31
Secretary of Defense, 47, 92, 101, 122, 124
security, vii, 1, 3, 4, 11, 15, 29, 51, 66, 67, 68, 73, 77, 79, 81, 82, 84, 88, 89, 90, 92, 94, 98, 106, 111, 113, 115, 126, 131, 143
security forces, 66, 77, 84, 106, 111
seizure, 65, 77
self-sufficiency, 15, 36, 38, 41
sellers, viii, 2, 36
Senate, 3, 46, 47, 48, 51, 52, 53, 54, 56, 57, 58, 59, 69, 76, 104, 123, 125, 141, 143
Senate Foreign Relations Committee, 46, 56
sensing, 33, 78
sensitivity, 131
sensors, 41
sentencing, 74, 75
separatism, 86, 112
September 11, 76, 92, 116, 117, 121, 137
services, 29, 47, 48, 132, 136
Shanghai Cooperation Organization, 113
shape, 21
Shiite factions, 108
Shiites, 103, 108
shoot, 120
shortage, 40
short-range ballistic missiles (SRBM), vii, 1
showing, 37, 83, 102
signs, ix, 37, 64, 111, 123, 132, 135
Singapore, 141
skilled personnel, 40
skimming, 110
Slovakia, 141
smuggling, 107
social behavior, 74
social problems, 132
socialization, 90
society, 86, 90, 134
software, 131, 132
solution, 42, 123, 124
Somalia, 116, 119
South Africa, 141
South America, 118
South Asia, 59, 113, 138
South Korea, 138, 141
sovereignty, 11, 65, 81, 105, 106, 115
Soviet Union, 5, 7, 9, 10, 25, 34, 37, 57, 58
space station, 34
speculation, 77
speech, 24, 56, 68, 76, 91, 92, 120, 121, 122, 124, 128, 129, 142
spending, 5
spin, 92
Spring, 142
Sri Lanka, 141
SS, 7, 21, 28, 102
stability, 19, 72, 115, 137
staffing, 135
state, viii, 2, 6, 20, 22, 24, 40, 47, 48, 53, 54, 67, 68, 72, 73, 74, 76, 79, 85, 86, 87, 89, 97, 101, 102, 103, 104, 105, 107, 111, 119, 122, 125, 126, 127, 136, 137, 141
state control, 67, 73, 74
state intervention, 73
State of the Union address, 3
statehood, 104
states, 3, 4, 5, 7, 8, 9, 40, 42, 47, 48, 51, 57, 79, 87, 90, 95, 103, 105, 107, 113, 126, 135, 136
steel, 138
stockpiling, 92
storage, 12, 81, 132
Strait of Hormuz, 44, 60, 88, 89, 90, 98, 106, 125
structure, 66, 90, 103

submarines, 37, 88, 124
Sudan, 116, 119
suicide, 78
Sunnis, 82, 85
supplier, 5, 88, 110
suppliers, 29, 38, 41
suppression, 77, 84, 131
Supreme Council, 30
surveillance, 114
survival, 4, 11
Sustainable Development, 117
Sweden, 87, 141
Switzerland, 142
sympathy, 109
Syria, ix, 15, 35, 36, 41, 48, 63, 64, 79, 90, 99, 103, 108, 109, 110, 111, 112, 123, 129, 132, 134, 135, 137, 141, 142

T

tactics, 13, 43, 114
Tajikistan, 113
takeover, 64, 77, 79, 109
Taliban, 114, 115, 116, 142
Tanzania, 116
target, 6, 8, 17, 29, 44, 45, 46, 51, 60, 76, 92, 102, 124
taxes, 68
teachers, 86
teams, 99
technical assistance, 20, 38, 56, 99, 141
techniques, 78
technological advances, 13
technologies, viii, 2, 8, 10, 27, 37, 39, 40, 47, 101, 131, 132
technology, viii, 2, 7, 17, 20, 21, 27, 28, 29, 32, 35, 36, 37, 38, 39, 40, 46, 47, 48, 54, 56, 57, 68, 88, 89, 90, 91, 94, 95, 97, 101, 102, 112, 115, 126, 129, 131, 132, 137
telecommunications, 32, 33
telephone, 100
television stations, 83
tensions, 106, 123
tenure, 108
territorial, 5, 51, 65, 111
territory, 5, 112
terrorism, 80, 82, 83, 84, 91, 104, 105, 109, 110, 111, 114, 124, 136
terrorist activities, 80
terrorist acts, 79
terrorist attack, 82
terrorist organization, 79, 80, 83, 103, 109
testing, vii, 1, 5, 7, 11, 15, 17, 36, 37, 41, 43, 44, 50, 56
textbook, 52
Thailand, 33, 105
Third World, 9, 52, 58
threat assessment, 12, 17, 36, 37, 38, 40, 41
threats, 5, 27, 47, 88, 98, 100, 107, 125
time frame, 27, 55, 94, 102, 111
torture, 65, 84, 86
tourism, 106
tracks, 43
trade, 7, 94, 101, 105, 106, 108, 118, 120, 121, 130, 132, 136, 137, 141
trafficking, 86, 113, 137
trafficking in persons, 86
training, 35, 37, 41, 43, 73, 89, 90, 109, 111, 114
trajectory, 18, 31, 51
transactions, ix, 64, 91, 106, 108, 113, 137
transcripts, 51, 57
translation, 11
transparency, 10
transport, 123, 136
Treasury, 47, 83, 91, 112, 114, 116, 136
treatment, 82, 85, 120
trial, 8, 86
triggers, 136
tumor, 67, 103
Turkey, 83, 88, 96, 97, 98, 102, 111, 123, 126, 133, 135, 138, 141
Turkmenistan, 64, 88, 133
turnout, 70, 75

U

U.N. Security Council, 47, 72, 76, 94, 95, 110, 113
U.S. assistance, 131, 132
U.S. policy, 3, 39, 81, 83, 120, 122, 123, 130, 133
UK, 8, 9, 24, 25, 28, 53, 56
Ukraine, 37, 39, 57, 88, 89
UN, 15, 18, 20, 22, 24, 25, 28, 29, 30, 41, 53, 54, 56, 59
UNHCR, 81, 82
unions, 85
United Kingdom, 57
United Nations, 54, 55, 56, 57, 59, 81, 83, 95, 106, 125
uranium, ix, 63, 89, 93, 94, 95, 96, 97, 99, 100, 101, 119, 122, 128

urban, 71, 73, 75, 77
urban youth, 73, 77
Uruguay, 117
USA, 51
USSR, 36
Uzbekistan, 113

V

variations, 39
vehicles, vii, 1, 7, 8, 27, 78, 141
Venezuela, 117, 118, 141
venue, 99
vessels, 44, 127
Vice President, 121
victims, 121
videos, 24
Vietnam, 141
violence, 79, 80, 82, 109, 119
vision, 90
Volkswagen, 138
vote, 70, 72, 73, 77, 78, 84, 85, 87, 90, 95, 108, 125, 136, 140, 141, 143
voters, 70
voting, 94, 140, 141
vulnerability, 78, 92, 106, 109, 110

W

war, viii, 2, 15, 36, 41, 45, 58, 60, 73, 88, 90, 107, 110, 111
Washington, 21, 45, 51, 52, 53, 55, 56, 57, 58, 59, 60, 61, 73, 102, 121, 131, 140, 141, 142, 143
water, 77, 89, 94, 95, 101, 118
weakness, 75
wealth, 107
weapon of mass destruction, 6
weapons, ix, 4, 10, 24, 38, 43, 47, 48, 54, 58, 60, 63, 90, 91, 92, 93, 100, 108, 109, 111, 112, 114, 119, 120, 136, 137, 141
weapons of mass destruction, 47, 120, 137
websites, 85, 98, 103, 133
West Bank, 90
Western countries, 57
White House, 51, 85, 123, 142, 143
wholesale, 138
Wisconsin, 53, 60
witchcraft, 74
withdrawal, 81, 110
WMD, 3, 4, 5, 6, 12, 37, 39, 40, 46, 47, 48, 54, 90, 91, 92, 95, 97, 101, 112, 125, 137
workers, 73
World Bank, 95
World Trade Organization, 94
World War I, 64
worldwide, 40, 90, 118, 136
worms, 34
worry, 55, 101, 124
wrestling, 115
WWW, 57

Y

Yale University, 131, 134
Yemen, 107, 141, 142
yield, 99, 104, 122, 138
young people, 129

Z

Zimbabwe, 119